W9-CHY-015

A handbook of

# Universal Conversion Factors

compiled and edited by

STEVEN GEROLDE



---

THE PETROLEUM PUBLISHING CO.

211 South Cheyenne
Tulsa, Oklahoma, USA

211 South Cheyenne
Tulsa, Oklahoma, U.S.A.

Library of Congress Catalog Card Number 71-164900

International Standard Book Number 87814-005-0

Printed in the United States of America

TO SUE AND LESLIE

# PREFACE

One of the most familiar and frequently recurring chores of the engineer, student, or technician is the rapid and exact conversion of measurement units of one system into the heterogeneous, yet related, units of other systems.

Each individual retains in his memory, or at easy access, the conversion factors of those units with which he is most familiar. If, however, he has to convert into units other than these, he must often refer to various handbooks before the desired development can be accomplished.

The conversion factors presented herein have resulted from many years of effort by the author to create one ready source, rather than many, for the conversion factors most often needed in his work. It is recognized that many conversion factors which one may find helpful are not included, but it is hoped that those most needed in every-day operations are included. It is also recognized that, while every effort has been made to establish the accuracy of each conversion factor, errors unknown to the author and to the publisher may exist. Both the author and the publisher welcome knowledge of any errors uncovered by those who use this handbook.

Steven Gerolde

# INTRODUCTION

The conversion factors presented herein are arranged alphabetically and are so listed in the table of contents to make it easy for the user of this handbook to readily locate the desired units. The conversion process for units listed under each heading and in the table of contents is simple. Just multiply the number of units corresponding to the proper heading by the conversion factor listed under that heading for the particular unit you wish to obtain or to which you wish to convert.

**Example.** Assume you wish to convert from acres to square miles, and that you have 1,000 acres. On Page 1 under acre, the conversion factor to convert to square miles is 0.0015625. Thus:

1,000 acres X 0.0015625 square miles per acre = 1.5625 square miles.

For conversion of those occasional units which do not have specific headings and are not listed in the table of contents, but which are listed under a heading is also relatively simple. Just divide the unit you wish to convert by the conversion factor for it. The result will be the unit corresponding to that heading.

**Example.** Assume you wish to convert from the unit, are (a metric unit of area equal to the area of a square 10 meters long on each side, or 100 square meters in area), to acres. The unit, are, is not listed by the various headings, but it is listed under the heading, acres, on Page 1. One acre equals 40.46875 ares. Let's assume you have 1,000 ares which you wish to convert to acres:

1,000 ares ÷ 40.46875 = 24.71 acres.

Similar conversions can be made for those occasional units given under the various headings but not set out by a main heading.

# CONTENTS

## ACRE: =

| | |
|---|---|
| 0.0015625 | square miles or sections |
| 0.004046875 | square kilometers |
| 0.1 | square furlongs |
| 0.4046875 | square hektometers |
| 10 | square chains |
| 40.46875 | square dekameters |
| 160 | square rods |
| 4,046.875 | square meters |
| 4,840 | square yards |
| 5,645.41213 | square varas (Texas) |
| 43,560 | square feet |
| 77,440 | square spans |
| 100,000 | square links |
| 400,000 | square hands |
| 404,687.5 | square decimeters |
| 40,468.750 | square centimeters |
| 6,272,640 | square inches |
| 4,046,875,000 | square millimeters |
| 0.4046875 | hectares |
| 40.46875 | ares |
| 4,046.875 | centares (centiares) |

## ACRE FOOT: =

| | |
|---|---|
| 1,233.48766 | kiloliters |
| 1,233.48766 | cubic meters |
| 1,633.3333 | cubic yards |
| 43,560 | cubic feet |
| 7,758.34 | barrels |
| 12,334.8766 | hektoliters |
| 35,000 | bushels—U.S. (dry) |
| 33,933.16195 | bushels—Imperial (dry) |
| 123,348.766 | dekaliters |
| 140,000 | pecks |
| 325,850.28 | gallons—U.S. (liquid) |
| 280,092.5925 | gallons—U.S. (dry) |
| 271,325.745265 | gallons—Imperial |
| 1,303,401.12 | quarts (liquid) |
| 1,120,370.370 | quarts (dry) |
| 1,233,487.66 | liters |
| 1,233,487.660 | cubic decimeters |
| 2,606,802.24 | pints |
| 10,427,208.96 | gills |

ACRE FOOT (cont'd): =

12,334,876.6 . . . . . . . . . . . . . . . . . . . . . . . . . . . . . . deciliters
75,271,680 . . . . . . . . . . . . . . . . . . . . . . . . . . . . . cubic inches
123,348,766 . . . . . . . . . . . . . . . . . . . . . . . . . . . . centiliters
1,233,487,660 . . . . . . . . . . . . . . . . . . . . . . . . . . milliliters
1,233,487,660 . . . . . . . . . . . . . . . . . . . . . . cubic centimeters
1,233,487,660,000 . . . . . . . . . . . . . . . . . . . .cubic millimeters

ACRE FOOT PER DAY: =

7,758.34 . . . . . . . . . . . . . . . . . . . . . . . . . . barrels per day
323.264167 . . . . . . . . . . . . . . . . . . . . . . . . barrels per hour
5.387736 . . . . . . . . . . . . . . . . . . . . . . . . barrels per minute
0.0897956 . . . . . . . . . . . . . . . . . . . . . . . barrels per second
325,850.28 . . . . . . . . . . . . . . . . . . . . . gallons (U.S.) per day
13,577.09400 . . . . . . . . . . . . . . . . . . . gallons (U.S.) per hour
226.284900 . . . . . . . . . . . . . . . . . . . gallons (U.S.) per minute
3.771415 . . . . . . . . . . . . . . . . . . . . gallons (U.S.) per second
271,325.745265 . . . . . . . . . . . . . . . .gallons (Imperial) per day
11,305.238400 . . . . . . . . . . . . . . . . gallons (Imperial) per hour
188.42066 . . . . . . . . . . . . . . . . . . .gallons (Imperial) per minute
3.140344 . . . . . . . . . . . . . . . . . . . .gallons (Imperial) per second
1,233.48766 . . . . . . . . . . . . . . . . . . . . . cubic meters per day
51.395319 . . . . . . . . . . . . . . . . . . . . . . cubic meters per hour
0.856589 . . . . . . . . . . . . . . . . . . . . . cubic meters per minute
0.0142765 . . . . . . . . . . . . . . . . . . . . .cubic meters per second
1,633.33333 . . . . . . . . . . . . . . . . . . . . . . cubic yards per day
68.0555552 . . . . . . . . . . . . . . . . . . . . . .cubic yards per hour
1.134259 . . . . . . . . . . . . . . . . . . . . . cubic yards per minute
0.0189043 . . . . . . . . . . . . . . . . . . . . cubic yards per second
43,560 . . . . . . . . . . . . . . . . . . . . . . . . . . cubic feet per day
1,815.0 . . . . . . . . . . . . . . . . . . . . . . . . . cubic feet per hour
30.25000 . . . . . . . . . . . . . . . . . . . . . . . cubic feet per minute
0.504167 . . . . . . . . . . . . . . . . . . . . . . . cubic feet per second
75,271,680.00 . . . . . . . . . . . . . . . . . . . .cubic inches per day
3,136,320.00 . . . . . . . . . . . . . . . . . . . . . cubic inches per hour
52,272.00 . . . . . . . . . . . . . . . . . . . . . .cubic inches per minute
871.200 . . . . . . . . . . . . . . . . . . . . . . . cubic inches per second

ATMOSPHERE: =

0.103327 . . . . . . . . . . . . . . . . . . . . .hektometers of water @ 60° F.
1.03327 . . . . . . . . . . . . . . . . . . . . . . dekameters of water @ 60° F.
10.3327 . . . . . . . . . . . . . . . . . . . . . . . . . meters of water @ 60° F.
33.9007 . . . . . . . . . . . . . . . . . . . . . . . . . . .feet of water @ 60° F.
406.8084 . . . . . . . . . . . . . . . . . . . . . . . . inches of water @ 60° F.

ATMOSPHERE (cont'd): =

103.327 . . . . . . . . . . . . . . . . . . . . . . . . . . . . .decimeters of water @ 60° F.
1,033.27 . . . . . . . . . . . . . . . . . . . . . . . . . . centimeters of water @ 60° F.
10,332.7 . . . . . . . . . . . . . . . . . . . . . . . . . . millimeters of water @ 60° F.
0.00760 . . . . . . . . . . . . . . . . . . . . . . . . . . hektometers of mercury @ 32° F.
0.0760 . . . . . . . . . . . . . . . . . . . . . . . . . .dekameters of mercury @ 32° F.
0.760 . . . . . . . . . . . . . . . . . . . . . . . . . . . . . meters of mercury @ 32° F.
2.49343 . . . . . . . . . . . . . . . . . . . . . . . . . . . . . feet of mercury @ 32° F.
29.9212 . . . . . . . . . . . . . . . . . . . . . . . . . . . .inches of mercury @ 32° F.
7.6 . . . . . . . . . . . . . . . . . . . . . . . . . . . . decimeters of mercury @ 32° F.
76 . . . . . . . . . . . . . . . . . . . . . . . . . . . centimeters of mercury @ 32° F.
760 . . . . . . . . . . . . . . . . . . . . . . . . . . .millimeters of mercury @ 32° F.
113,893.88 . . . . . . . . . . . . . . . . . . . . . . . .tons per square hektometer
1,138.9388 . . . . . . . . . . . . . . . . . . . . . . . . tons per square dekameter
11.389388 . . . . . . . . . . . . . . . . . . . . . . . . . . . tons per square meter
1.0581 . . . . . . . . . . . . . . . . . . . . . . . . . . . . .tons per square foot
0.00734792 . . . . . . . . . . . . . . . . . . . . . . . . . . tons per square inch
0.11389388 . . . . . . . . . . . . . . . . . . . . . . .tons per square decimeter
0.0011389388 . . . . . . . . . . . . . . . . . . . . . . tons per square centimeter
0.000011389388 . . . . . . . . . . . . . . . . . . . . tons per square millimeter
103,327,000 . . . . . . . . . . . . . . . . . . . . kilograms per square hektometer
1,033,270 . . . . . . . . . . . . . . . . . . . . . . kilograms per square dekameter
10,332.7 . . . . . . . . . . . . . . . . . . . . . . . . . kilograms per square meter
959.931252 . . . . . . . . . . . . . . . . . . . . . . . .kilograms per square foot
6.666189 . . . . . . . . . . . . . . . . . . . . . . . . . .kilograms per square inch
103.327 . . . . . . . . . . . . . . . . . . . . . . . kilograms per square decimeter
1.03327 . . . . . . . . . . . . . . . . . . . . . . . kilograms per square centimeter
0.0103327 . . . . . . . . . . . . . . . . . . . . . . kilograms per square millimeter
227,774,851.2 . . . . . . . . . . . . . . . . . . . . .pounds per square hektometer
2,277,748.512 . . . . . . . . . . . . . . . . . . . . pounds per square dekameter
22,777.48512 . . . . . . . . . . . . . . . . . . . . . . . . pounds per square meter
2,116.080 . . . . . . . . . . . . . . . . . . . . . . . . . . pounds per square foot
14.696 . . . . . . . . . . . . . . . . . . . . . . . . . . . . pounds per square inch
227.7748512 . . . . . . . . . . . . . . . . . . . . . .pounds per square decimeter
2.277748512 . . . . . . . . . . . . . . . . . . . . . pounds per square centimeter
0.02277748512 . . . . . . . . . . . . . . . . . . . .pounds per square millimeter
1,033,270,000 . . . . . . . . . . . . . . . . . . hektograms per square hektometer
10,332,700 . . . . . . . . . . . . . . . . . . . . hektograms per square dekameter
103,327 . . . . . . . . . . . . . . . . . . . . . . . . .hektograms per square meter
9,599,31252 . . . . . . . . . . . . . . . . . . . . . . .hektograms per square foot
66.66189 . . . . . . . . . . . . . . . . . . . . . . . . .hektograms per square inch
1,033.27 . . . . . . . . . . . . . . . . . . . . . . hektograms per square decimeter
10.3327 . . . . . . . . . . . . . . . . . . . . . . . hektograms per square centimeter
0.103327 . . . . . . . . . . . . . . . . . . . . . . hektograms per square millimeter
10,332,700,000 . . . . . . . . . . . . . . . . . . . dekagrams per square hektometer

ATMOSPHERE (cont'd): =

| | |
|---|---|
| 103,327,000 | dekagrams per square dekameter |
| 1,033,270 | dekagrams per square meter |
| 95,993.1252 | dekagrams per square foot |
| 666.6189 | dekagrams per square inch |
| 10,332.7 | dekagrams per square decimeter |
| 103.327 | dekagrams per square centimeter |
| 1.03327 | dekagrams per square millimeter |
| 3,644,397,619.2 | ounces per square hektometer |
| 36,443,976.192 | ounces per square dekameter |
| 364,439.76192 | ounces per square meter |
| 33,857.28 | ounces per square foot |
| 235.136 | ounces per square inch |
| 3,644.39762 | ounces per square decimeter |
| 36.44398 | ounces per square centimeter |
| 0.36444 | ounces per square millimeter |
| 103,327,000,000 | grams per square hektometer |
| 1,033,270,000 | grams per square dekameter |
| 10,332,700 | grams per square meter |
| 959,931.252 | grams per square foot |
| 6,666,189 | grams per square inch |
| 103,327 | grams per square decimeter |
| 1,033.27 | grams per square centimeter |
| 10.3327 | grams per square millimeter |
| 1,033,270,000,000 | decigrams per square hektometer |
| 10,332,700,000 | decigrams per square dekameter |
| 103,327,000 | decigrams per square meter |
| 9,599,312.52 | decigrams per square foot |
| 66,661.89 | decigrams per square inch |
| 1,033,270 | decigrams per square decimeter |
| 10,332.7 | decigrams per square centimeter |
| 103.327 | decigrams per square millimeter |
| 10,332,700,000,000 | centigrams per square hektometer |
| 103,327,000,000 | centigrams per square dekameter |
| 1,033,270,000 | centigrams per square meter |
| 95,933,125.2 | centigrams per square foot |
| 666,618.9 | centigrams per square inch |
| 10,332,700 | centigrams per square decimeter |
| 103,327 | centigrams per square centimeter |
| 1,033.270 | centigrams per square millimeter |
| 103,327,000,000,000 | milligrams per square hektometer |
| 1,033,270,000,000 | milligrams per square dekameter |
| 10,332,700,000 | milligrams per square meter |
| 959,931,252 | milligrams per square foot |
| 6,666,189 | milligrams per square inch |

**ATMOSPHERE (cont'd): =**

| | |
|---|---|
| 103,327,000 | milligrams per square decimeter |
| 1,033,270 | milligrams per square centimeter |
| 10,332.7 | milligrams per square millimeter |
| 101,325,000,000,000 | dynes per square hektometer |
| 1,013,250,000,000 | dynes per square dekameter |
| 10,132,500,000 | dynes per square meter |
| 941,343,587 | dynes per square foot |
| 6,537,096 | dynes per square inch |
| 101,325,000 | dynes per square decimeter |
| 1,013,250 | dynes per square centimeter |
| 10,132.5 | dynes per square millimeter |
| 1.01325 | bars |

**BARREL: =**

| | |
|---|---|
| 0.158987 | kiloliters |
| 0.158987 | cubic meters |
| 0.20794 | cubic yards |
| 1.58987 | hectoliters |
| 4.511274 | bushels—U.S. (dry) |
| 4.373766 | bushels—Imperial (dry) |
| 5.6146 | cubic feet |
| 15.89871 | dekaliters |
| 18.045097 | pecks |
| 42 | gallons—U.S. (liquid) |
| 36.09798 | gallons—U.S. (dry) |
| 34.99089 | gallons—Imperial |
| 168 | quarts (liquid) |
| 144.408516 | quarts (dry) |
| 158.987146 | liters |
| 158.987146 | cubic decimeters |
| 336 | pints |
| 1,344 | gills |
| 1,589.87146 | deciliters |
| 9,702.0288 | cubic inches |
| 15,898.71459456 | centiliters |
| 158,987.1459456 | milliliters |
| 158,987.1459456 | cubic centimeters |
| 158,987,145.9456 | cubic millimeters |
| 0.174993 | tons (short) of water @ 62° F. |
| 0.1562438 | tons (long) of water @ 62° F. |
| 0.1587512 | tons (metric) of water @ 62° F. |
| 158.7512 | kilograms of water @ 62° F. |
| 349.986 | pounds of water @ 62° F. |

BARREL (cont'd): =

15.87512 . . . . . . . . . . . . . . . . . . . . . . . . hektograms of water @ 62° F.
1.587512 . . . . . . . . . . . . . . . . . . . . . . . . .dekagrams of water @ 62° F.
5,599.776 . . . . . . . . . . . . . . . . . . . . . . . . .ounces of water @ 62° F.
0.1587512 . . . . . . . . . . . . . . . . . . . . . . . . grams of water @ 62° F.
0.01587512 . . . . . . . . . . . . . . . . . . . . . . . decigrams of water @ 62° F.
0.001587512 . . . . . . . . . . . . . . . . . . . . . . centigrams of water @ 62° F.
0.0001587512 . . . . . . . . . . . . . . . . . . . . . milligrams of water @ 62° F.
5.1042 . . . . . . . . . . . . . . . . . . . . . . . . . . . . . . sacks of cement
2,449,902 . . . . . . . . . . . . . . . . . . . . . . . . . . . . . . . . . . . . grains
404.25 . . . . . . . . . . . . . . . . . . . . . . . .pounds of salt water @ 60° F. of 1.155 specific gravity

BARREL OF CEMENT: =

0.158987 . . . . . . . . . . . . . . . . . . . . . . . . . . . . . . . . . . . kiloliters
0.158987 . . . . . . . . . . . . . . . . . . . . . . . . . . . . . . . . cubic meters
0.20794 . . . . . . . . . . . . . . . . . . . . . . . . . . . . . . . . . cubic yards
1.58987 . . . . . . . . . . . . . . . . . . . . . . . . . . . . . . . . . hectoliters
4.511274 . . . . . . . . . . . . . . . . . . . . . . . . . . . . bushels—U.S. (dry)
4.373766 . . . . . . . . . . . . . . . . . . . . . . . . . . bushels—Imperial (dry)
5.6146 . . . . . . . . . . . . . . . . . . . . . . . . . . . . . . . . . . .cubic feet
15.89871 . . . . . . . . . . . . . . . . . . . . . . . . . . . . . . . . . dekaliters
18.045097 . . . . . . . . . . . . . . . . . . . . . . . . . . . . . . . . . . . . pecks
42 . . . . . . . . . . . . . . . . . . . . . . . . . . . . . . . gallons—U.S. (liquid)
36.10213 . . . . . . . . . . . . . . . . . . . . . . . . . . . . . . .gallons—U.S. (dry)
34.99089 . . . . . . . . . . . . . . . . . . . . . . . . . . . . . . .gallons—Imperial
168 . . . . . . . . . . . . . . . . . . . . . . . . . . . . . . . . . . . quarts (liquid)
144.408516 . . . . . . . . . . . . . . . . . . . . . . . . . . . . . . . .quarts (dry)
158.987146 . . . . . . . . . . . . . . . . . . . . . . . . . . . . . . . . . . . . .liters
158.987146 . . . . . . . . . . . . . . . . . . . . . . . . . . . . . cubic decimeters
336 . . . . . . . . . . . . . . . . . . . . . . . . . . . . . . . . . . . . . . . . . .pints
1,344 . . . . . . . . . . . . . . . . . . . . . . . . . . . . . . . . . . . . . . . . . gills
1,589.87146 . . . . . . . . . . . . . . . . . . . . . . . . . . . . . . . . . deciliters
9,702.0288 . . . . . . . . . . . . . . . . . . . . . . . . . . . . . . . . cubic inches
15,898.71459456 . . . . . . . . . . . . . . . . . . . . . . . . . . . . . .centiliters
158,987.1459456 . . . . . . . . . . . . . . . . . . . . . . . . . . . . . . milliliters
158,987.1459456 . . . . . . . . . . . . . . . . . . . . . . . . . cubic centimeters
158,987,145.9456 . . . . . . . . . . . . . . . . . . . . . . . . .cubic millimeters
0.188 . . . . . . . . . . . . . . . . . . . . . . . . . . . . . . . . . . . .tons (short)
0.16796 . . . . . . . . . . . . . . . . . . . . . . . . . . . . . . . . . . tons (long)
0.170551 . . . . . . . . . . . . . . . . . . . . . . . . . . . . . . . . tons (metric)
170.55097 . . . . . . . . . . . . . . . . . . . . . . . . . . . . . . . . . . kilograms
376 . . . . . . . . . . . . . . . . . . . . . . . . . . . . . . . . . . . . . . . . pounds
1705.5097 . . . . . . . . . . . . . . . . . . . . . . . . . . . . . . . . . .hektograms

**BARREL OF CEMENT: (cont'd): =**

| | |
|---|---|
| 17,055.097 | dekagrams |
| 6,016 | ounces |
| 170,550.97 | grams |
| 1,705,509.7 | decigrams |
| 17,055,097 | centigrams |
| 170,550,970 | milligrams |

**BARREL PER DAY: =**

| | |
|---|---|
| 0.041667 | barrels per hour |
| 0.00069444 | barrels per minute |
| 0.000011574 | barrels per second |
| 0.1589871 | kiloliters per day |
| 0.0066245 | kiloliters per hour |
| 0.00011041 | kiloliters per minute |
| 0.000001840 | kiloliters per second |
| 0.1589871 | cubic meters per day |
| 0.0066245 | cubic meters per hour |
| 0.00011041 | cubic meters per minute |
| 0.000001840 | cubic meters per second |
| 0.20794 | cubic yards per day |
| 0.0086642 | cubic yards per hour |
| 0.0001444 | cubic yards per minute |
| 0.0000024067 | cubic yards per second |
| 1.589871 | hektoliters per day |
| 0.066245 | hektoliters per hour |
| 0.0011041 | hektoliters per minute |
| 0.00001840 | hektoliters per second |
| 5.6146 | cubic feet per day |
| 0.233942 | cubic feet per hour |
| 0.00389903 | cubic feet per minute |
| 0.0000649838 | cubic feet per second |
| 15.89871 | dekaliters per day |
| 0.66245 | dekaliters per hour |
| 0.011041 | dekaliters per minute |
| 0.0001840 | dekaliters per second |
| 42 | gallons (U.S.) per day |
| 1.71875 | gallons (U.S.) per hour |
| 0.029167 | gallons (U.S.) per minute |
| 0.0004861 | gallons (U.S.) per second |
| 34.99089 | gallons (Imperial) per day |
| 1.45795 | gallons (Imperial) per hour |
| 0.024299 | gallons (Imperial) per minute |
| 0.00040499 | gallons (Imperial) per second |

**BARREL PER DAY (cont'd): =**

| | |
|---|---|
| 168 | quarts (U.S.) per day |
| 6.875 | quarts (U.S.) per hour |
| 0.11668 | quarts (U.S.) per minute |
| 0.0019444 | quarts (U.S.) per second |
| 158.98714 | liters per day |
| 6.6245 | liters per hour |
| 0.11041 | liters per minute |
| 0.001840 | liters per second |
| 158.98714 | cubic decimeters per day |
| 6.6245 | cubic decimeters per hour |
| 0.11041 | cubic decimeters per minute |
| 0.001840 | cubic decimeters per second |
| 336 | pints per day |
| 13.75 | pints per hour |
| 0.23336 | pints per minute |
| 0.0038888 | pints per second |
| 1,344 | gills per day |
| 55 | gills per hour |
| 0.933344 | gills per minute |
| 0.0155552 | gills per second |
| 1,589.87146 | deciliters per day |
| 66.245 | deciliters per hour |
| 1.1041 | deciliters per minute |
| 0.01840 | deciliters per second |
| 9,702.0288 | cubic inches per day |
| 404.2 | cubic inches per hour |
| 6.7375 | cubic inches per minute |
| 0.112292 | cubic inches per second |
| 15,898.7146 | centiliters per day |
| 662.45 | centiliters per hour |
| 11.041 | centiliters per minute |
| 0.1840 | centiliters per second |
| 158,987.145946 | milliliters per day |
| 6,624.5 | milliliters per hour |
| 110.41 | milliliters per minute |
| 1.84 | milliliters per second |
| 158,987.145946 | cubic centimeters per day |
| 6,624.5 | cubic centimeters per hour |
| 110.41 | cubic centimeters per minute |
| 1.840 | cubic centimeters per second |
| 158,987,145.946 | cubic millimeters per day |
| 6,624,500 | cubic millimeters per hour |
| 110,410 | cubic millimeters per minute |
| 1,840 | cubic millimeters per second |

## BARREL PER HOUR: =

| | |
|---|---|
| 24 | barrels per day |
| 0.016667 | barrels per minute |
| 0.000277778 | barrels per second |
| 3.81567 | kiloliters per day |
| 0.1589871 | kiloliters per hour |
| 0.0026498 | kiloliters per minute |
| 0.000044163 | kiloliters per second |
| 3.81567 | cubic meters per day |
| 0.1589871 | cubic meters per hour |
| 0.0026498 | cubic meters per minute |
| 0.000044163 | cubic meters per second |
| 4.99056 | cubic yards per day |
| 0.20794 | cubic yards per hour |
| 0.0034657 | cubic yards per minute |
| 0.000057761 | cubic yards per second |
| 38.1567 | hektoliters per day |
| 1.589871 | hektoliters per hour |
| 0.026498 | hektoliters per minute |
| 0.00044163 | hektoliters per second |
| 134.7504 | cubic feet per day |
| 5.6146 | cubic feet per hour |
| 0.093577 | cubic feet per minute |
| 0.15596 | cubic feet per second |
| 381.567 | dekaliters per day |
| 15.89871 | dekaliters per hour |
| 0.26498 | dekaliters per minute |
| 0.0044163 | dekaliters per second |
| 1,008 | gallons (U.S.) per day |
| 42 | gallons (U.S.) per hour |
| 0.7 | gallons (U.S.) per minute |
| 0.11667 | gallons (U.S.) per second |
| 839.78136 | gallons (Imperial) per day |
| 34.99089 | gallons (Imperial) per hour |
| 0.58318 | gallons (Imperial) per minute |
| 0.0097197 | gallons (Imperial) per second |
| 4,032 | quarts (U.S.) per day |
| 168 | quarts (U.S.) per hour |
| 2.80 | quarts (U.S.) per minute |
| 0.046667 | quarts (U.S.) per second |
| 3,815.6904 | liters per day |
| 158.98714 | liters per hour |
| 2.64979 | liters per minute |
| 0.044163 | liters per second |
| 3,815.6904 | cubic decimeters per day |
| 158.98714 | cubic decimeters per hour |

**BARREL PER HOUR (cont'd):=**

| | |
|---|---|
| 2.64979 | cubic decimeters per minute |
| 0.044163 | cubic decimeters per second |
| 8,064 | pints per day |
| 336 | pints per hour |
| 5.60 | pints per minute |
| 0.93333 | pints per second |
| 32,256 | gills per day |
| 1,344 | gills per hour |
| 22.40 | gills per minute |
| 0.37333 | gills per second |
| 38,156.904 | deciliters per day |
| 1,589.87146 | deciliters per hour |
| 26.49786 | deciliters per minute |
| 0.44163 | deciliters per second |
| 232,848 | cubic inches per day |
| 9,702.0288 | cubic inches per hour |
| 161.7014 | cubic inches per minute |
| 2.695 | cubic inches per second |
| 381,569.04 | centiliters per day |
| 15,898.7146 | centiliters per hour |
| 264.97858 | centiliters per minute |
| 4.4163 | centiliters per second |
| 3,815,690.4 | milliliters per day |
| 158,987.145946 | milliliters per hour |
| 2,649.78576 | milliliters per minute |
| 44.163 | milliliters per second |
| 3,815,690.4 | cubic centimeters per day |
| 158,987.145946 | cubic centimeters per hour |
| 2,649.78576 | cubic centimeters per minute |
| 44.163 | cubic centimeters per second |
| 3,815,690,400.0 | cubic millimeters per day |
| 158,987,145,946 | cubic millimeters per hour |
| 2,649,785.76 | cubic millimeters per minute |
| 44,163.096 | cubic millimeters per second |

**BARREL PER MINUTE: =**

| | |
|---|---|
| 1,440 | barrels per day |
| 60 | barrels per hour |
| 0.016667 | barrels per second |
| 228.94272 | kiloliters per day |
| 9.53928 | kiloliters per hour |
| 0.158987 | kiloliters per minute |

BARREL PER MINUTE (cont'd):=

| | |
|---|---|
| 0.0026498 | kiloliters per second |
| 228.94272 | cubic meters per day |
| 9.53928 | cubic meters per hour |
| 0.158987 | cubic meters per minute |
| 0.0026498 | cubic meters per second |
| 299.43648 | cubic yards per day |
| 12.47652 | cubic yards per hour |
| 0.20794 | cubic yards per minute |
| 0.0034657 | cubic yards per second |
| 2,289.4272 | hektoliters per day |
| 95.3928 | hektoliters per hour |
| 1.58987 | hektoliters per minute |
| 0.026498 | hektoliters per second |
| 8,085.05280 | cubic feet per day |
| 336.8772 | cubic feet per hour |
| 5.6146 | cubic feet per minute |
| 0.093577 | cubic feet per second |
| 22,894.272 | dekaliters per day |
| 953.928 | dekaliters per hour |
| 15.8987 | dekaliters per minute |
| 0.26498 | dekaliters per second |
| 60,480 | gallons (U.S.) per day |
| 2,520 | gallons (U.S.) per hour |
| 42 | gallons (U.S.) per minute |
| 0.7 | gallons (U.S.) per second |
| 50,386.7520 | gallons (Imperial) per day |
| 2,099.4480 | gallons (Imperial) per hour |
| 34.99089 | gallons (Imperial) per minute |
| 0.58318 | gallons (Imperial) per second |
| 241,920 | quarts per day (U.S.) |
| 10,080 | quarts per hour (U.S.) |
| 168 | quarts per minute (U.S.) |
| 2.80 | quarts per second (U.S.) |
| 228,941.48966 | liters per day |
| 9,539.22874 | liters per hour |
| 158.987146 | liters per minute |
| 2.64979 | liters per second |
| 228,941.48966 | cubic decimeters per day |
| 9,539.22874 | cubic decimeters per hour |
| 158.987146 | cubic decimeters per minute |
| 2.64979 | cubic decimeters per second |
| 483,840 | pints per day |
| 20,160 | pints per hour |
| 336 | pints per minute |

**BARREL PER MINUTE (cont'd):=**

5.60 . . . . . . . . pints per second
1,935,360 . . . . . . . . gills per day
80,640 . . . . . . . . gills per hour
1,344 . . . . . . . . gills per minute
22.40 . . . . . . . . gills per second
2,289,414.89664 . . . . . . . . deciliters per day
95,392.28736 . . . . . . . . deciliters per hour
1,589.87146 . . . . . . . . deciliters per minute
26.49786 . . . . . . . . deciliters per second
13,970,921.472 . . . . . . . . cubic inches per day
582,121.7280 . . . . . . . . cubic inches per hour
9,702.0288 . . . . . . . . cubic inches per minute
161.70048 . . . . . . . . cubic inches per second
22,894,148.9664 . . . . . . . . centiliters per day
953,922.8736 . . . . . . . . centiliters per hour
15,898.71459 . . . . . . . . centiliters per minute
264.97858 . . . . . . . . centiliters per second
228,941,489.664 . . . . . . . . milliliters per day
9,539,228.736 . . . . . . . . milliliters per hour
158,987.14595 . . . . . . . . milliliters per minute
2,649.78576 . . . . . . . . milliliters per second
228,941,489.664 . . . . . . . . cubic centimeters per day
9,539,228.736 . . . . . . . . cubic centimeters per hour
158,987.14595 . . . . . . . . cubic centimeters per minute
2,649.78576 . . . . . . . . cubic centimeters per second
228,941,489,664 . . . . . . . . cubic millimeters per day
9,539,228,736 . . . . . . . . cubic millimeters per hour
158,987,145.946 . . . . . . . . cubic millimeters per minute
2,649,785.76 . . . . . . . . cubic millimeters per second

**BARREL PER SECOND:=**

86,400 . . . . . . . . barrels per day
3,600 . . . . . . . . barrels per hour
60 . . . . . . . . barrels per minute
13,736.47680 . . . . . . . . kiloliters per day
572.35320 . . . . . . . . kiloliters per hour
9.53922 . . . . . . . . kiloliters per minute
0.158987 . . . . . . . . kiloliters per second
13,736.47680 . . . . . . . . cubic meters per day
572.35320 . . . . . . . . cubic meters per hour
9.53922 . . . . . . . . cubic meters per minute
0.158987 . . . . . . . . cubic meters per second

BARREL PER SECOND (cont'd):=

17,966.0160 . . . . . . . . . . cubic yards per day
748.5840 . . . . . . . . . . .cubic yards per hour
12.47640 . . . . . . . . . . cubic yards per minute
0.20794 . . . . . . . . . . cubic yards per second
137,364.7680 . . . . . . . . . . .hektoliters per day
5,723.5320 . . . . . . . . . . hektoliters per hour
95.3922 . . . . . . . . . . .hektoliters per minute
1.58987 . . . . . . . . . . .hektoliters per second
485,101.44 . . . . . . . . . . cubic feet per day
20,212.560 . . . . . . . . . . cubic feet per hour
336.8760 . . . . . . . . . . cubic feet per minute
5.6146 . . . . . . . . . . cubic feet per second
1,373,648.5440 . . . . . . . . . . dekaliters per day
57,235,356 . . . . . . . . . . .dekaliters per hour
953.92260 . . . . . . . . . . dekaliters per minute
15.89871 . . . . . . . . . . dekaliters per second
3,628,800 . . . . . . . . . . gallons (U.S.) per day
151,200 . . . . . . . . . . gallons (U.S.) per hour
2,520 . . . . . . . . . . gallons (U.S.) per minute
42 . . . . . . . . . . gallons (U.S.) per second
3,023,212.8960 . . . . . . . . . . .gallons (Imperial) per day
125,967.2040 . . . . . . . . . . gallons (Imperial) per hour
2,099.45340 . . . . . . . . . . .gallons (Imperial) per minute
34.99089 . . . . . . . . . . .gallons (Imperial) per second
14,515,200 . . . . . . . . . . quarts (U.S.) per day
604,800 . . . . . . . . . . .quarts (U.S.) per hour
10,080 . . . . . . . . . . quarts (U.S.) per minute
168 . . . . . . . . . . quarts (U.S.) per second
13,736,489.4144 . . . . . . . . . . liters per day
572,353.7256 . . . . . . . . . . liters per hour
9,539.22876 . . . . . . . . . . liters per minute
158.987146 . . . . . . . . . . liters per second
13,736,489.4144 . . . . . . . . . . cubic decimeters per day
572,353.7256 . . . . . . . . . . cubic decimeters per hour
9,539.22876 . . . . . . . . . . cubic decimeters per minute
158.987146 . . . . . . . . . . cubic decimeters per second
29,030,400 . . . . . . . . . . pints per day
1,209,600 . . . . . . . . . . pints per hour
20,160 . . . . . . . . . . pints per minute
336 . . . . . . . . . . pints per second
116,121,600 . . . . . . . . . . .gills per day
4,838,400 . . . . . . . . . . gills per hour
80,640 . . . . . . . . . . .gills per minute
1,344 . . . . . . . . . . .gills per second
137,364,894.144 . . . . . . . . . . .deciliters per day

BARREL PER SECOND (cont'd):=

5,723,537.256 . . . . . . . . . . . . . . . . . . . . . . . . . deciliters per hour
95,392.28754 . . . . . . . . . . . . . . . . . . . . . . . . . deciliters per minute
1,589.87146 . . . . . . . . . . . . . . . . . . . . . . . . . deciliters per second
838,255,288.320 . . . . . . . . . . . . . . . . . . . . . . cubic inches per day
34,927,303.6800 . . . . . . . . . . . . . . . . . . . . . . cubic inches per hour
582,131.7280 . . . . . . . . . . . . . . . . . . . . . . . cubic inches per minute
9,702.0288 . . . . . . . . . . . . . . . . . . . . . . . . cubic inches per second
1,373,648,941.440 . . . . . . . . . . . . . . . . . . . . . centiliters per day
57,235,372.56 . . . . . . . . . . . . . . . . . . . . . . . centiliters per hour
953,922.8754 . . . . . . . . . . . . . . . . . . . . . . . centiliters per minute
15,898.71459 . . . . . . . . . . . . . . . . . . . . . . . centiliters per second
13,736,489,414.40 . . . . . . . . . . . . . . . . . . . . . milliliters per day
572,353,725.6 . . . . . . . . . . . . . . . . . . . . . . . milliliters per hour
9,539,228.760 . . . . . . . . . . . . . . . . . . . . . . . milliliters per minute
158,987.14595 . . . . . . . . . . . . . . . . . . . . . . . milliliters per second
13,736,489,414.4 . . . . . . . . . . . . . . . . . . . . cubic centimeters per day
572,353,725.6 . . . . . . . . . . . . . . . . . . . . . . cubic centimeters per hour
9,539,228.760 . . . . . . . . . . . . . . . . . . . . . cubic centimeters per minute
158,987.14595 . . . . . . . . . . . . . . . . . . . . . cubic centimeters per second
13,736,489,414,400 . . . . . . . . . . . . . . . . . . . cubic millimeters per day
572,353,725,600 . . . . . . . . . . . . . . . . . . . . . cubic millimeters per hour
9,539,228,754 . . . . . . . . . . . . . . . . . . . . . cubic millimeters per minute
158,987,145.9456 . . . . . . . . . . . . . . . . . . . . cubic millimeters per second

BTU (60° F.): =

25,030 . . . . . . . . . . . . . . . . . . . . . . . . . . . . . . foot poundals
300,360 . . . . . . . . . . . . . . . . . . . . . . . . . . . . . inch poundals
777.97265 . . . . . . . . . . . . . . . . . . . . . . . . . . . . foot pounds
9,335.67120 . . . . . . . . . . . . . . . . . . . . . . . . . . . inch pounds
0.00027776 . . . . . . . . . . . . . . . . . . . . . . . ton (short) calories
0.25198 . . . . . . . . . . . . . . . . . . . . . . . . . . . kilogram calories
0.55552 . . . . . . . . . . . . . . . . . . . . . . . . . . . . pound calories
2.5198 . . . . . . . . . . . . . . . . . . . . . . . . . . . hektogram calories
25.198 . . . . . . . . . . . . . . . . . . . . . . . . . . . dekagram calories
8.88832 . . . . . . . . . . . . . . . . . . . . . . . . . . . . . ounce calories
251.98 . . . . . . . . . . . . . . . . . . . . . . . . . . . . . . gram calories
2,519.8 . . . . . . . . . . . . . . . . . . . . . . . . . . . . decigram calories
25,198 . . . . . . . . . . . . . . . . . . . . . . . . . . . . centigram calories
251.980 . . . . . . . . . . . . . . . . . . . . . . . . . . . . milligram calories
0.000012201 . . . . . . . . . . . . . . . . . . . . . . . . . . . kilowatt days
0.00029283 . . . . . . . . . . . . . . . . . . . . . . . . . . . kilowatt hours
0.01757 . . . . . . . . . . . . . . . . . . . . . . . . . . . . kilowatt minutes

BTU (60° F.) (cont'd):=

| | |
|---|---|
| 1.0546 | kilowatt seconds |
| 0.012201 | watt days |
| 0.29283 | watt hours |
| 17.57 | watt minutes |
| 1,054.6 | watt seconds |
| 0.11856 | ton meters |
| 107.56 | kilogram meters |
| 237.12678 | pound meters |
| 1,075.6 | hektogram meters |
| 10,756 | dekagram meters |
| 3,794.02848 | ounce meters |
| 107,560 | gram meters |
| 1,075,600 | decigram meters |
| 10,756,000 | centigram meters |
| 107,560,000 | milligram meters |
| 0.0011856 | ton hektometers |
| 1.0756 | kilogram hektometers |
| 2.37127 | pound hektometers |
| 10.756 | hektogram hektometers |
| 107.56 | dekagram hektometers |
| 37.94028 | ounce hektometers |
| 1,075.6 | gram hektometers |
| 10,756 | decigram hektometers |
| 107,560 | centigram hektometers |
| 1,075,600 | milligram hektometers |
| 0.011856 | ton dekameters |
| 10.756 | kilogram dekameters |
| 23.7127 | pound dekameters |
| 107.56 | hektogram dekameters |
| 1,075.6 | dekagram dekameters |
| 379.40285 | ounce dekameters |
| 10,756 | gram dekameters |
| 107,560 | decigram dekameters |
| 1,075,600 | centigram dekameters |
| 10,756,000 | milligram dekameters |
| 0.388977 | ton feet |
| 352.887473 | kilogram feet |
| 777.97265 | pound feet |
| 3,528.874731 | hektogram feet |
| 35,288.747308 | dekagram feet |
| 12,447.611780 | ounce feet |
| 352,887.473080 | gram feet |
| 3,528.875 | decigram feet |
| 35,288,747 | centigram feet |

BTU (60° F.) (cont'd):=

| | |
|---|---|
| 352,887,473 | milligram feet |
| 4.667724 | ton inches |
| 4,234.649677 | kilogram inches |
| 9,335.671800 | pound inches |
| 42,346.496772 | hektogram inches |
| 423,464.96772 | dekagram inches |
| 149,371.34136 | ounce inches |
| 4,234,650 | gram inches |
| 42,346,497 | decigram inches |
| 423,464,968 | centigram inches |
| 4,234,649,677 | milligram inches |
| 1,1856 | ton decimeters |
| 1.0756 | kilogram decimeters |
| 2,371.2678 | pound decimeters |
| 10,756 | hektogram decimeters |
| 107,560 | dekagram decimeters |
| 37,940.2848 | ounce decimeters |
| 1,075,600 | gram decimeters |
| 10,756,000 | decigram decimeters |
| 107,560,000 | centigram decimeters |
| 1,075,600,000 | milligram decimeters |
| 11.856 | ton centimeters |
| 10,756 | kilogram centimeters |
| 23,712.678 | pound centimeters |
| 107,560 | hektogram centimeters |
| 1,075,600 | dekagram centimeters |
| 379,402.848 | ounce centimeters |
| 10,756,000 | gram centimeters |
| 107,560,000 | decigram centimeters |
| 1,075,600,000 | centigram centimeters |
| 10,756,000,000 | milligram centimeters |
| 118.56 | ton millimeters |
| 107,560 | kilogram millimeters |
| 237,126.780 | pound millimeters |
| 1,075,600 | hektogram millimeters |
| 10,756,000 | dekagram millimeters |
| 3,794,028.48 | ounce millimeters |
| 107,560,000 | gram millimeters |
| 1,075,600,000 | decigram millimeters |
| 10,756,000,000 | centigram millimeters |
| 107,560,000,000 | milligram millimeters |
| 0.0104028 | kiloliter-atmospheres |
| 0.104028 | hektoliter-atmospheres |
| 1.040277 | dekaliter-atmospheres |
| 10.40277 | liter-atmospheres |

**BTU (60° F.) (cont'd):=**

| | |
|---|---|
| 104.0277 | deciliter-atmospheres |
| 1,040.277 | centiliter-atmospheres |
| 10,402.77 | milliliter-atmospheres |
| 0.0000000000104104 | cubic kilometer-atmospheres |
| 0.0000000104104 | cubic hektometer-atmospheres |
| 0.000010410 | cubic dekameter-atmospheres |
| 0.0104104 | cubic meter-atmospheres |
| 0.3676637 | cubic feet-atmospheres |
| 635.277597 | cubic inch-atmospheres |
| 10.410432 | cubic decimeter-atmospheres |
| 10,410.4320 | cubic centimeter-atmospheres |
| 10,410,432 | cubic millimeter-atmospheres |
| 1,054.198 | joules |
| 0.0003982 | Cheval-vapeur hours |
| 0.000016372 | horsepower days |
| 0.00039292 | horsepower hours |
| 0.0235757 | horsepower minutes |
| 1.41451 | horsepower seconds |
| 0.0000685 | pounds of carbon oxidized with perfect efficiency |
| 0.001030 | pounds of water evaporated from and at 212° F. |

**BTU (60° F.) PER DAY: =**

| | |
|---|---|
| 25,030 | foot poundals per day |
| 1,042.92 | foot poundals per hour |
| 17.3820 | foot poundals per minute |
| 0.2897 | foot poundals per second |
| 777.97265 | foot pounds per day |
| 32.41553 | foot pounds per hour |
| 0.54026 | foot pounds per minute |
| 0.0090043 | foot pounds per second |
| 0.25198 | kilogram calories per day |
| 0.010499 | kilogram calories per hour |
| 0.00017498 | kilogram calories per minute |
| 0.0000029164 | kilogram calories per second |
| 8.88832 | ounce calories per day |
| 0.37035 | ounce calories per hour |
| 0.0061724 | ounce calories per minute |
| 0.00010287 | ounce calories per second |
| 107.56 | kilogram meters per day |
| 4.48164 | kilogram meters per hour |

BTU ($60^{\circ}$ F.) PER DAY:=

| | |
|---|---|
| 0.074694 | kilogram meters per minute |
| 0.0012449 | kilogram meters per second |
| 10.40277 | liter-atmosphere per day |
| 0.43344 | liter-atmospheres per hour |
| 0.007224 | liter-atmospheres per minute |
| 0.00012040 | liter-stmospheres per second |
| 0.3676637 | cubic foot-atmospheres per day |
| 0.0153166 | cubic foot-atmospheres per hour |
| 0.000255276 | cubic foot-atmospheres per minute |
| 0.0000042546 | cubic foot-atmospheres per second |
| 0.000016148 | Cheval-Vapeurs |
| 0.000016372 | horsepower |
| 0.000012201 | kilowatts |
| 0.012201 | watts |
| 1,054.198 | joules per day |
| 43.9236 | joules per hour |
| 0.73206 | joules per minute |
| 0.012201 | joules per second |
| 0.0000685 | pounds of carbon oxidized with perfect efficiency per day |
| 0.00000285415 | pounds of carbon oxidized with perfect efficiency per hour |
| 0.000000047569 | pounds of carbon oxidized with perfect efficiency per minute |
| 0.00000000079282 | pounds of carbon oxidized with perfect efficiency per second |
| 0.001030 | pounds of water evaporated from and at $212^{\circ}$ F. per day |
| 0.000042916 | pounds of water evaporated from and at $212^{\circ}$ F. per hour |
| 0.00000071526 | pounds of water evaporated from and at $212^{\circ}$ F. per minute |
| 0.000000011921 | pounds of water evaporated from and at $212^{\circ}$ F. per second |
| 1.0 | BTU per day |
| 0.0416667 | BTU per hour |
| 0.00069444 | BTU per minute |
| 0.000011574 | BTU per second |

BTU PER HOUR: =

| | |
|---|---|
| 600,720.1920 | foot poundals per day |
| 25,030 | foot poundals per hour |

**BTU PER HOUR (cot'd):=**

| | |
|---|---|
| 417.16680 | foot poundals per minute |
| 6.95278 | foot poundals per second |
| 18,671.34359 | foot pounds per day |
| 777.97265 | foot pounds per hour |
| 12.96621 | foot pounds per minute |
| 0.21610 | foot pounds per second |
| 6.04748 | kilogram calories per day |
| 0.25198 | kilogram calories per hour |
| 0.0041996 | kilogram calories per minute |
| 0.000069994 | kilogram calories per second |
| 213.31968 | ounce calories per day |
| 8.88832 | ounce calories per hour |
| 0.14814 | ounce calories per minute |
| 0.0024690 | ounce calories per second |
| 2,581.4592 | kilogram meters per day |
| 107.56 | kilogram meters per hour |
| 1.79268 | kilogram meters per minute |
| 0.029878 | kilogram meters per second |
| 249.67008 | liter-atmospheres per day |
| 10.40277 | liter-atmospheres per hour |
| 0.173382 | liter-atmospheres per minute |
| 0.0028897 | liter-atmospheres per second |
| 8.822304 | cubic foot-atmospheres per day |
| 0.3676637 | cubic foot-atmospheres per hour |
| 0.00612660 | cubic foot-atmospheres per minute |
| 0.00010211 | cubic foot-atmospheres per second |
| 0.00038754 | Cheval-vapeurs |
| 0.00039292 | horsepower |
| 0.00029283 | kilowatts |
| 0.29283 | watts |
| 25,300.5120 | joules per day |
| 1,054.198 | joules per hour |
| 17.569967 | joules per minute |
| 0.292833 | joules per second |
| 0.00164402 | pounds of carbon oxidized with perfect efficiency per day |
| 0.0000685 | pounds of carbon oxidized with perfect efficiency per hour |
| 0.0000011417 | pounds of carbon oxidized with perfect efficiency per minute |
| 0.000000019028 | pounds of carbon oxidized with perfect efficiency per second |
| 0.02472 | pounds of water evaporated from and at 212° F. per day |
| 0.001030 | pounds of water evaporated from and at 212° F. per hour |

**BTU PER HOUR (cont'd):=**

|  |  |
|---|---|
| 0.000017167 | pounds of water evaporated from and at 212° F. per minute |
| 0.00000028611 | pounds of water evaporated from and at 212° F. per second |
| 24 | BTU per day |
| 1.0 | BTU per hour |
| 0.016667 | BTU per minute |
| 0.00027778 | BTU per second |

**BTU PER MINUTE: =**

|  |  |
|---|---|
| 36,043,200 | foot poundals per day |
| 1,501,800 | foot poundals per hour |
| 25,030 | foot poundals per minute |
| 417.16667 | foot poundals per second |
| 1,120,281 | foot pounds per day |
| 46,678.35899 | foot pounds per hour |
| 777.97265 | foot pounds per minute |
| 12.96621 | foot pounds per second |
| 362.854 | kilogram calories per day |
| 15.11892 | kilogram calories per hour |
| 0.25198 | kilogram calories per minute |
| 0.0041997 | kilogram calories per second |
| 12,799.180803 | ounce calories per day |
| 533.299200 | ounce calories per hour |
| 8.88832 | ounce calories per minute |
| 0.148139 | ounce calories per second |
| 154,886.688000 | kilogram meters per day |
| 6,453.6120 | kilogram meters per hour |
| 107.56 | kilogram meters per minute |
| 1.79267 | kilogram meters per second |
| 14,980.0320 | liter-atmospheres per day |
| 624.168 | liter-atmospheres per hour |
| 10.40277 | liter-atmospheres per minute |
| 0.17338 | liter-atmospheres per second |
| 529.435728 | cubic foot-atmospheres per day |
| 22.059822 | cubic foot-atmospheres per hour |
| 0.3676637 | cubic foot-atmospheres per minute |
| 0.00612773 | cubic foot-atmospheres per second |
| 0.023252 | Cheval-vapeur hours |
| 0.023575 | horsepower |
| 0.01757 | kilowatts |
| 17.57 | watts |

**BTU PER MINUTE (cont'd):=**

1,518,048 . . . . . . . . . . . . . . . . . . . . . . . . . . . . . joules per day
63,252 . . . . . . . . . . . . . . . . . . . . . . . . . . . . . . . . joules per hour
1,054.198 . . . . . . . . . . . . . . . . . . . . . . . . . . . . joules per minute
17.569967 . . . . . . . . . . . . . . . . . . . . . . . . . . . .joules per second
0.098643 . . . . . . . . . . . . . . . pounds of carbon oxidized with perfect efficiency per day
0.0041101 . . . . . . . . . . . . . . pounds of carbon oxidized with perfect efficiency per hour
0.0000685 . . . . . . . . . . . . . . pounds of carbon oxidized with perfect efficiency per minute
0.0000011417 . . . . . . . . . . . . pounds of carbon oxidized with perfect efficiency per second
1.48320 . . . . . . . . . . . . . . . . . . .pounds of water evaporated from and at 212° F. per day
0.06180 . . . . . . . . . . . . . . . . . . .pounds of water evaporated from and at 212° F. hour
0.001030 . . . . . . . . . . . . . . . . . .pounds of water evaporated from and at 212° F. per minute
0.000017167 . . . . . . . . . . . . . . . .pounds of water evaporated from and at 212° F. per second
1,440 . . . . . . . . . . . . . . . . . . . . . . . . . . . . . . . . . BTU per day
60 . . . . . . . . . . . . . . . . . . . . . . . . . . . . . . . . . . BTU per hour
1.0 . . . . . . . . . . . . . . . . . . . . . . . . . . . . . . . . BTU per minute
0.016667 . . . . . . . . . . . . . . . . . . . . . . . . . . . . . BTU per second

**BTU PER SECOND:=**

2,162,592,000 . . . . . . . . . . . . . . . . . . . . . . . .foot poundals per day
90,108,000 . . . . . . . . . . . . . . . . . . . . . . . . . . foot poundals per hour
1,501,800 . . . . . . . . . . . . . . . . . . . . . . . . .foot poundals per minute
25,030 . . . . . . . . . . . . . . . . . . . . . . . . . . .foot poundals per second
67,216,836.960 . . . . . . . . . . . . . . . . . . . . . . . .foot pounds per day
2,800,701.540 . . . . . . . . . . . . . . . . . . . . . . . . foot pounds per hour
46,678.3590 . . . . . . . . . . . . . . . . . . . . . . . .foot pounds per minute
777.97265 . . . . . . . . . . . . . . . . . . . . . . . . . .foot pounds per second
21,771.0720 . . . . . . . . . . . . . . . . . . . . . . kilogram calories per day
907.1280 . . . . . . . . . . . . . . . . . . . . . . . . kilogram calories per hour
15.1188 . . . . . . . . . . . . . . . . . . . . . . . . kilogram calories per minute
0.25198 . . . . . . . . . . . . . . . . . . . . . . . . kilogram calories per second
767,950.8480 . . . . . . . . . . . . . . . . . . . . . . . .ounce calories per day
31,997.9520 . . . . . . . . . . . . . . . . . . . . . . . . ounce calories per hour
533.29920 . . . . . . . . . . . . . . . . . . . . . . . .ounce calories per minute
8.88832 . . . . . . . . . . . . . . . . . . . . . . . . . .ounce calories per second
9,293,184.0 . . . . . . . . . . . . . . . . . . . . . . . kilogram meters per day

BTU PER SECOND (cont'd):=

387,216.0 . . . . . . . . . . kilogram meters per hour
6,453.60 . . . . . . . . . . kilogram meters per minute
107.56 . . . . . . . . . . kilogram meters per second
898,799.3280 . . . . . . . . . . liter-atmospheres per day
37,449.9720 . . . . . . . . . . liter-atmospheres per hour
624.16620 . . . . . . . . . . liter-atmospheres per minute
10.40277 . . . . . . . . . . liter-atmospheres per second
31,766.143680 . . . . . . . . . . cubic foot-atmospheres per day
1,323.589320 . . . . . . . . . . cubic foot-atmospheres per hour
22.0598220 . . . . . . . . . . cubic foot-atmospheres per minute
0.3676637 . . . . . . . . . . cubic foot-atmospheres per second
1.39519 . . . . . . . . . . Cheval-vapeurs
1,41454 . . . . . . . . . . horsepower
1.055 . . . . . . . . . . kilowatts
1,055 . . . . . . . . . . watts
91,082,707.2 . . . . . . . . . . joules per day
3,795,112.8 . . . . . . . . . . joules per hour
63,251.880 . . . . . . . . . . joules per minute
1,054.198 . . . . . . . . . . joules per second
5.9184 . . . . . . . . . . pounds of carbon oxidized with perfect efficiency per day
0.2466 . . . . . . . . . . pounds of carbon oxidized with perfect efficiency per hour
0.00411 . . . . . . . . . . pounds of carbon oxidized with perfect efficiency per minute
0.0000685 . . . . . . . . . . pounds of carbon oxidized with perfect efficiency per second
88.992000 . . . . . . . . . . pounds of water evaporated from and at 212° F. per day
3.70800 . . . . . . . . . . pounds of water evaporated from and at 212° F. per hour
0.061800 . . . . . . . . . . pounds of water evaporated from and at 212° F. per minute
0.001030 . . . . . . . . . . pounds of water evaporated from and at 212° F. per second
86,400 . . . . . . . . . . BTU per day
3,600 . . . . . . . . . . BTU per hour
60 . . . . . . . . . . BTU per minute
1 . . . . . . . . . . BTU per second

BTU PER SQUARE FOOT PER DAY: =

23,505.12 . . . . . . . . . . kilowatts per square hektometer
235.0512 . . . . . . . . . . kilowatts per square dekameter

## BTU PER SQUARE FOOT PER DAY:=

| | |
|---|---|
| 2,350512 | kilowatts per square meter |
| 25.30051 | kilowatts per square foot |
| 0.17569 | kilowatts per square inch |
| 0.023505 | kilowatts per square decimeter |
| 0.00023505 | kilowatts per square centimeter |
| 0.0000023505 | kilowatts per square millimeter |
| 23,505,120 | watts per square hektometer |
| 235,051.2 | watts per square dekameter |
| 2,350.512 | watts per square meter |
| 25,300.512 | watts per square foot |
| 175.6944 | watts per square inch |
| 23.50512 | watts per square decimeter |
| 0.23505 | watts per square centimeter |
| 0.0023502 | watts per square millimeter |
| 31,537.728 | horsepower per square hektometer |
| 315.37728 | horsepower per square dekameter |
| 3.15377 | horsepower per square meter |
| 33.94829 | horsepower per square foot |
| 0.23576 | horsepower per square inch |
| 0.031538 | horsepower per square decimeter |
| 0.00031538 | horsepower per square centimeter |
| 0.0000031538 | horsepower per square millimeter |

## BTU PER SQUARE FOOT PER HOUR: =

| | |
|---|---|
| 979.380 | kilowatts per square hektometer |
| 9.7938 | kilowatts per square dekameter |
| 0.097938 | kilowatts per square meter |
| 1.05419 | kilowatts per square foot |
| 0.0073206 | kilowatts per square inch |
| 0.000979380 | kilowatts per square decimeter |
| 0.00000979380 | kilowatts per square centimeter |
| 0.0000000979380 | kilowatts per square millimeter |
| 97,938 | watts per square hektometer |
| 9,793.800 | watts per square dekameter |
| 97.93800 | watts per square meter |
| 1,054.188 | watts per square foot |
| 7.3206 | watts per square inch |
| 0.979380 | watts per square decimeter |
| 0.0097938 | watts per square centimeter |
| 0.000097938 | watts per square millimeter |
| 1,314.0720 | horsepower per square hektometer |
| 13.14072 | horsepower per square dekameter |

## BTU PER SQUARE FOOT PER HOUR (cont'd):=

| | |
|---|---|
| 0.13141 | horsepower per square meter |
| 1.41451 | horsepower per square foot |
| 0.00982332 | horsepower per square inch |
| 0.00131407 | horsepower per square decimeter |
| 0.000013141 | horsepower per square centimeter |
| 0.00000013141 | horsepower per square millimeter |

## BTU PER SQUARE FOOT PER MINUTE: =

| | |
|---|---|
| 16.323 | kilowatts per square hektometer |
| 0.16323 | kilowatts per square dekamter |
| 0.0016323 | kilowatts per square meter |
| 0.01757 | kilowatts per square foot |
| 0.00012201 | kilowatts per square inch |
| 0.000016323 | kilowatts per square decimeter |
| 0.00000016323 | kilowatts per square centimeter |
| 0.0000000016323 | kilowatts per square millimeter |
| 16,323 | watts per square hektometer |
| 163.23 | watts per square dekameter |
| 1.6323 | watts per square meter |
| 17.57 | watts per square foot |
| 0.12201 | watts per square inch |
| 0.016323 | watts per square decimeter |
| 0.00016323 | watts per square centimeter |
| 0.0000016323 | watts per square millimeter |
| 21.90118 | horsepower per square hektometer |
| 0.21901 | horsepower per square dekameter |
| 0.0021901 | horsepower per square meter |
| 0.023575 | horsepower per square foot |
| 0.00016372 | horsepower per square inch |
| 0.000021901 | horsepower per square decimeter |
| 0.00000021901 | horsepower per square centimeter |
| 0.0000000021901 | horsepower per square millimeter |

## BTU PER SQUARE FOOT PER SECOND: =

| | |
|---|---|
| 0.27205 | kilowatts per square hektometer |
| 0.0027205 | kilowatts per square dekameter |
| 0.000027205 | kilowatts per square meter |
| 0.00029283 | kilowatts per square foot |
| 0.0000020335 | kilowatts per square inch |
| 0.00000027205 | kilowatts per square decimeter |

## BTU PER SQUARE FOOT PER SECOND:=

| | |
|---|---|
| 0.0000000027205 | kilowatts per square centimeter |
| 0.000000000027205 | kilowatts per square millimeter |
| 272.05 | watts per square hektometer |
| 2.7205 | watts per square dekameter |
| 0.027205 | watts per square meter |
| 0.29283 | watts per square foot |
| 0.0020335 | watts per square inch |
| 0.00027205 | watts per square decimeter |
| 0.0000027205 | watts per square centimeter |
| 0.000000027205 | watts per square millimeter |
| 0.36502 | horsepower per square hektometer |
| 0.0036502 | horsepower per square dekameter |
| 0.000036502 | horsepower per square meter |
| 0.00039292 | horsepower per square foot |
| 0.0000027287 | horsepower per square inch |
| 0.00000036502 | horsepower per square decimeter |
| 0.0000000036502 | horsepower per square centimeter |
| 0.000000000036502 | horsepower per square millimeter |

## BUSHELS—U.S. (DRY): =

| | |
|---|---|
| 0.035238 | kiloliters |
| 0.035238 | cubic meters |
| 0.04609 | cubic yards |
| 0.304785 | barrels—U.S. |
| 0.35238 | hectoliters |
| 0.96945 | bushels—Imp. (dry) |
| 1.24446 | cubic feet |
| 3.5238 | dekaliters |
| 4 | pecks |
| 9.3088 | gallons—U.S. (liquid) |
| 8 | gallons—U.S. (dry) |
| 7.81457 | gallons—Imp. |
| 37.2353 | quarts (liquid) |
| 32 | quarts (dry) |
| 35.238 | liters |
| 35.238 | cubic decimeters |
| 64 | pints (dry) |
| 74.8706 | pints (liquid) |
| 299.4824 | gills (liquid) |
| 352.38 | deciliters |
| 2,150.42 | cubic inches |
| 3,523.8 | centiliters |
| 35,238 | millimeters |

BUSHEL–U.S. (DRY):=

| | |
|---|---|
| 35,238 | cubic centimeters |
| 35,238,000 | cubic millimeters |
| 0.053335 | tons (short) |
| 0.047621 | tons (long) |
| 0.048385 | tons (metric) |
| 48.38492 | kilograms |
| 106.67048 | pounds |
| 483.84924 | hektograms |
| 4,838.4924 | dekagrams |
| 7,741.58787 | ounces |
| 48,384.924 | grams |
| 483,849.24 | decigrams |
| 4,838,492.4 | centigrams |
| 48,384,924 | milligrams |

BUSHEL–IMPERIAL: =

| | |
|---|---|
| 0.036348 | kiloliters |
| 0.036348 | cubic meters |
| 0.047542 | cubic yards |
| 0.31439 | barrels |
| 0.36348 | hectoliters |
| 1.03151 | bushels–U.S. |
| 1.2843 | cubic feet |
| 3.63484 | dekaliters |
| 4.12604 | pecks |
| 9.60212 | gallons–U.S. (liquid) |
| 8.25208 | gallons–U.S. (dry) |
| 8 | gallons–Imp. |
| 38.40858 | quarts (liquid) |
| 33.00832 | quarts (dry) |
| 36.34835 | liters |
| 36.34835 | cubic decimeters |
| 66.01664 | pints (dry) |
| 76.81716 | pints (liquid) |
| 307.26856 | gills (liquid) |
| 363.4835 | deciliters |
| 2,219.3 | cubic inches |
| 3,634.835 | centiliters |
| 36,348.35 | milliliters |
| 36,348.35 | cubic centimeters |
| 36,348,350 | cubic millimeters |
| 0.055016 | tons (short) |

**BUSHEL—IMPERIAL:=**

| | |
|---|---|
| 0.049122 | tons (long) |
| 0.049910 | tons (metric) |
| 49.90953 | kilograms |
| 110.031667 | pounds |
| 499.095330 | hektograms |
| 4,990.95330 | dekagrams |
| 7,985.52530 | ounces |
| 49,909.53296 | grams |
| 499,095.32955 | decigrams |
| 4,990,953.29552 | centigrams |
| 49,909,532.95524 | milligrams |

**CENTARE (CENTIARE): =**

| | |
|---|---|
| 0.0000003831 | square miles or sections |
| 0.0000001111 | square kilometers |
| 0.000024710 | square furlongs |
| 0.00024710 | acres |
| 0.0001111 | square hektometers |
| 0.00247104 | square chains |
| 0.01111 | square dekameters |
| 0.039537 | square rods |
| 1 | square meters |
| 1.19598 | square yards |
| 1.39498 | square varas (Texas) |
| 10.7639 | square feet |
| 19.13580 | square spans |
| 27.7104 | square links |
| 96.8750 | square hands |
| 100 | square decimeters |
| 10,000 | square centimeters |
| 1,550 | square inches |
| 1,000,000 | square millimeters |
| 0.0001 | hectares |
| 0.01 | ares |

**CENTIGRAM: =**

| | |
|---|---|
| 0.0000000110231 | tons (short) |
| 0.00000000984206 | tons (long) |
| 0.00000001 | tons (metric) |
| 0.00001 | kilograms |

## CENTIGRAM (cont'd):=

| | |
|---|---|
| 0.0000267923 | pounds (Troy) |
| 0.000022406 | pounds (Avoir) |
| 0.0001 | hektograms |
| 0.001 | dekagrams |
| 0.000321507 | ounces (Troy) |
| 0.000352739 | ounces (Avoir) |
| 0.01 | grams |
| 0.1 | decigrams |
| 10 | milligrams |
| 0.1543236 | grains |
| 0.00257206 | drachmas (fluid) |
| 0.00257206 | drams (Troy) |
| 0.0056438 | drams (Avoir) |
| 0.006430149 | pennyweight |
| 0.00771618 | scruples |
| 0.05 | carats (metric) |

## CENTILITERS: =

| | |
|---|---|
| 0.000001 | kiloliters |
| 0.00001 | cubic meters |
| 0.00001308 | cubic yards |
| 0.000062897 | barrels |
| 0.0001 | hectoliters |
| 0.00028377 | bushels—U.S. (dry) |
| 0.00027510 | bushels—Imp. (dry) |
| 0.00035314 | cubic feet |
| 0.001 | dekaliters |
| 0.0026417 | gallons—U.S. (liquid) |
| 0.0022707 | gallons—U.S. (dry) |
| 0.0021997 | gallons—Imp. |
| 0.010567 | quarts (liquid) |
| 0.0090828 | quarts (dry) |
| 0.01 | liters |
| 0.01 | cubic decimeters |
| 0.018161 | pints |
| 0.072663 | gills |
| 0.1 | deciliters |
| 0.61025 | cubic inches |
| 10 | milliliters |
| 10 | cubic centimeters |
| 10,000 | cubic millimeters |
| 0.33815 | ounces (fluid) |
| 2.70518 | drams (fluid) |

**CENTIMETER: =**

| | |
|---|---|
| 0.0000062137 | miles |
| 0.00001 | kilometers |
| 0.000049709 | furlongs |
| 0.0001 | hektometers |
| 0.00049709 | chains |
| 0.001 | dekameters |
| 0.0019884 | rods |
| 0.01 | meters |
| 0.010936 | yards |
| 0.011811 | varas (Texas) |
| 0.032808 | feet |
| 0.043744 | spans |
| 0.049709 | links |
| 0.098424 | hands |
| 0.1 | decimeters |
| 0.3937 | inches |
| 1.00 | centimeters |
| 10 | millimeters |
| 393.70 | mils |
| 10,000 | microns |
| 10,000,000 | milli-microns |
| 10,000,000 | micro-millimeters |
| 100,000,000 | Angstrom units |
| 15,531.6 | wave lengths of red line of cadmium |

**CENTIMETERS PER DAY: =**

| | |
|---|---|
| 0.000006214 | miles per day |
| 0.00000025892 | miles per hour |
| 0.0000000043153 | miles per minute |
| 0.000000000071921 | miles per second |
| 0.00001 | kilometers per day |
| 0.00000041667 | kilometers per hour |
| 0.0000000069444 | kilometers per minute |
| 0.00000000011574 | kilometers per second |
| 0.000049709 | furlongs per day |
| 0.0000020712 | furlongs per hour |
| 0.000000034520 | furlongs per minute |
| 0.00000000057534 | furlongs per second |
| 0.0001 | hektometers per day |
| 0.0000041667 | hektometers per hour |
| 0.000000069444 | hektometers per minute |
| 0.0000000011574 | hektometers per second |

CENTIMETERS PER DAY (cont'd):=

| | |
|---|---|
| 0.00049709 | chains per day |
| 0.000020712 | chains per hour |
| 0.00000034520 | chains per minute |
| 0.0000000057534 | chains per second |
| 0.001 | dekameters per day |
| 0.000041667 | dekameters per hour |
| 0.00000069444 | dekameters per minute |
| 0.000000011574 | dekameters per second |
| 0.0019884 | rods per day |
| 0.000082850 | rods per hour |
| 0.0000013808 | rods per minute |
| 0.000000023014 | rods per second |
| 0.01 | meters per day |
| 0.00041667 | meters per hour |
| 0.0000069444 | meters per minute |
| 0.00000011574 | meters per second |
| 0.010936 | yards per day |
| 0.00045667 | yards per hour |
| 0.0000075944 | yards per minute |
| 0.00000012657 | yards per second |
| 0.011811 | varas (Texas) per day |
| 0.00049212 | varas (Texas) per hour |
| 0.000008202 | varas (Texas) per minute |
| 0.00000013670 | varas (Texas) per second |
| 0.032808 | feet per day |
| 0.0013670 | feet per hour |
| 0.000022783 | feet per minute |
| 0.00000037972 | feet per second |
| 0.043744 | spans per day |
| 0.0018227 | spans per hour |
| 0.000030378 | spans per minute |
| 0.00000050630 | spans per second |
| 0.049709 | links per day |
| 0.0020712 | links per hour |
| 0.000034520 | links per minute |
| 0.00000057534 | links per second |
| 0.098424 | hands per day |
| 0.0041010 | hands per hour |
| 0.000068350 | hands per minute |
| 0.0000011392 | hands per second |
| 0.1 | decimeters per day |
| 0.0041667 | decimeters per hour |
| 0.000069444 | decimeters per minute |
| 0.0000011574 | decimeters per second |
| 1 | centimeters per day |

## CENTIMETERS PER DAY (cont'd):=

| | |
|---|---|
| 0.041667 | centimeters per hour |
| 0.00069444 | centimeters per minute |
| 0.000011574 | centimeters per second |
| 0.3937 | inches per day |
| 0.016404 | inches per hour |
| 0.00027340 | inches per minute |
| 0.0000045567 | inches per second |
| 10 | millimeters per day |
| 0.41667 | millimeters per hour |
| 0.0069444 | millimeters per minute |
| 0.00011574 | millimeters per second |
| 393.70 | mils per day |
| 16.40417 | mils per hour |
| 0.27340 | mils per minute |
| 0.0045567 | mils per second |
| 10,000 | microns per day |
| 416.66667 | microns per hour |
| 6.9444 | microns per minute |
| 0.11574 | microns per second |

## CENTIMETERS PER HOUR: =

| | |
|---|---|
| 0.00014914 | miles per day |
| 0.000006214 | miles per hour |
| 0.00000010357 | miles per minute |
| 0.0000000017261 | miles per second |
| 0.00024000 | kilometers per day |
| 0.00001 | kilometers per hour |
| 0.00000016667 | kilometers per minute |
| 0.0000000027778 | kilometers per second |
| 0.0011930 | furlongs per day |
| 0.000049709 | furlongs per hour |
| 0.00000082848 | furlongs per minute |
| 0.000000013808 | furlongs per second |
| 0.0024000 | hektometers per day |
| 0.0001 | hektometers per hour |
| 0.0000016667 | hectometers per minute |
| 0.000000027778 | hektometers per second |
| 0.011930 | chains per day |
| 0.00049709 | chains per hour |
| 0.0000082848 | chains per minute |
| 0.00000013808 | chains per second |
| 0.024000 | dekameters per day |
| 0.001 | dekameters per hour |

**CENTIMETERS PER HOUR (cont'd): =**

| | |
|---|---|
| 0.000016667 | dekameters per minute |
| 0.00000027778 | dekameters per second |
| 0.047722 | rods per day |
| 0.0019884 | rods per hour |
| 0.000033140 | rods per minute |
| 0.00000055233 | rods per second |
| 0.24000 | meters per day |
| 0.01 | meters per hour |
| 0.00016667 | meters per minute |
| 0.0000027778 | meters per second |
| 0.262464 | yards per day |
| 0.010936 | yards per hour |
| 0.00018227 | yards per minute |
| 0.0000030378 | yards per second |
| 0.28346 | varas (Texas) per day |
| 0.011811 | varas (Texas) per hour |
| 0.00019685 | varas (Texas) per minute |
| 0.0000032808 | varas (Texas) per second |
| 0.78739 | feet per day |
| 0.032808 | feet per hour |
| 0.00054680 | feet per minute |
| 0.0000091133 | feet per second |
| 1.049856 | spans per day |
| 0.043744 | spans per hour |
| 0.00072907 | spans per minute |
| 0.000012151 | spans per second |
| 1.19302 | links per day |
| 0.049709 | links per hour |
| 0.00082848 | links per minute |
| 0.000013808 | links per second |
| 2.36218 | hands per day |
| 0.098424 | hands per hour |
| 0.0016404 | hands per minute |
| 0.000027340 | hands per second |
| 2.40000 | decimeters per day |
| 0.1 | decimeters per hour |
| 0.0016667 | decimeters per minute |
| 0.000027778 | decimeters per second |
| 24.00000 | centimeters per day |
| 1 | centimeters per hour |
| 0.016667 | centimeters per minute |
| 0.00027778 | centimeters per second |
| 9.44880 | inches per day |
| 0.3937 | inches per hour |
| 0.0065617 | inches per minute |

CENTIMETERS PER HOUR (cont'd): =

| | |
|---|---|
| 0.00010936 | inches per second |
| 240.00000 | millimeters per day |
| 10 | millimeters per hour |
| 0.16667 | millimeters per minute |
| 0.0027778 | millimeters per second |
| 9,448.8 | mils per day |
| 393.70 | mils per hour |
| 6.56167 | mils per minute |
| 0.109361 | mils per second |
| 240,000 | microns per day |
| 10,000 | microns per hour |
| 166.66667 | microns per minute |
| 2.77778 | microns per second |

CENTIMETER PER MINUTE: =

| | |
|---|---|
| 0.0089482 | miles per day |
| 0.00037284 | miles per hour |
| 0.000006214 | miles per minute |
| 0.00000010357 | miles per second |
| 0.014400 | kilometers per day |
| 0.00060000 | kilometers per hour |
| 0.00001 | kilometers per minute |
| 0.00000016667 | kilometers per second |
| 0.071581 | furlongs per day |
| 0.0029825 | furlongs per hour |
| 0.000049709 | furlongs per minute |
| 0.00000082848 | furlongs per second |
| 0.14400 | hektometers per day |
| 0.0060000 | hektometers per hour |
| 0.0001 | hektometers per minute |
| 0.0000016667 | hektometers per second |
| 0.71581 | chains per day |
| 0.029825 | chains per hour |
| 0.00049709 | chains per minute |
| 0.0000082848 | chains per second |
| 1.44000 | dekameters per day |
| 0.060000 | dekameters per hour |
| 0.001 | dekameters per minute |
| 0.000016667 | dekameters per second |
| 2.86330 | rods per day |
| 0.11930 | rods per hour |
| 0.0019884 | rods per minute |

CENTIMETER PER MINUTE (cont'd): =

| | |
|---|---|
| 0.000033140 | rods per second |
| 14.40000 | meters per day |
| 0.60000 | meters per hour |
| 0.01 | meters per minute |
| 0.00016667 | meters per second |
| 15.74784 | yards per day |
| 0.656160 | yards per hour |
| 0.010936 | yards per minute |
| 0.00018227 | yards per second |
| 17.00784 | varas (Texas) per day |
| 0.70866 | varas (Texas) per hour |
| 0.011811 | varas (Texas) per minute |
| 0.00019685 | varas (Texas) per second |
| 47.24352 | feet per day |
| 1.96848 | feet per hour |
| 0.032808 | feet per minute |
| 0.00054680 | feet per second |
| 62.99136 | spans per day |
| 2.62464 | spans per hour |
| 0.043744 | spans per minute |
| 0.00072907 | spans per second |
| 71.58096 | links per day |
| 2.98254 | links per hour |
| 0.049709 | links per minute |
| 0.00082848 | links per second |
| 141.73056 | hands per day |
| 5.90544 | hands per hour |
| 0.098424 | hands per minute |
| 0.0016404 | hands per second |
| 144.00 | decimeters per day |
| 6.00 | decimeters per hour |
| 0.1 | decimeters per minute |
| 0.0016667 | decimeters per second |
| 1,440.00 | centimeters per day |
| 60.00 | centimeters per hour |
| 1.0 | centimeters per minute |
| 0.016667 | centimeters per second |
| 566.92800 | inches per day |
| 23.62200 | inches per hour |
| 0.3937 | inches per minute |
| 0.0065617 | inches per second |
| 14,440.00 | millimeters per day |
| 600.00 | millimeters per hour |
| 10 | millimeters per minute |

**CENTIMETER PER MINUTE (cont'd): =**

| | |
|---|---|
| 0.16667 | millimeters per second |
| 566,928 | mils per day |
| 23,622 | mils per hour |
| 393.70 | mils per minute |
| 6.56167 | mils per second |
| 14,440,000 | microns per day |
| 600,000 | microns per hour |
| 10,000 | microns per minute |
| 166.66667 | microns per second |

**CENTIMETER PER SECOND: =**

| | |
|---|---|
| 0.53689 | miles per day |
| 0.022370 | miles per hour |
| 0.00037284 | miles per minute |
| 0.000006214 | miles per second |
| 0.8640 | kilometers per day |
| 0.0360 | kilometers per hour |
| 0.0006 | kilometers per minute |
| 0.00001 | kilometers per second |
| 4.29486 | furlongs per day |
| 0.17895 | furlongs per hour |
| 0.0029825 | furlongs per minute |
| 0.000049709 | furlongs per second |
| 8.640 | hektometers per day |
| 0.360 | hektometers per hour |
| 0.006 | hektometers per minute |
| 0.0001 | hektometers per second |
| 42.94858 | chains per day |
| 1.78952 | chains per hour |
| 0.029825 | chains per minute |
| 0.00049709 | chains per second |
| 86.400 | dekameters per day |
| 3.600 | dekameters per hour |
| 0.060 | dekameters per minute |
| 0.001 | dekameters per second |
| 171.79776 | rods per day |
| 7.15824 | rods per hour |
| 0.11930 | rods per minute |
| 0.0019884 | rods per second |
| 864.0 | meters per day |
| 36.0 | meters per hour |
| 0.60 | meters per minute |

CENTIMETER PER SECOND (cont'd): =

| | |
|---|---|
| 0.01 | meters per second |
| 944.8704 | yards per day |
| 39.36960 | yards per hour |
| 0.656160 | yards per minute |
| 0.010936 | yards per second |
| 1,020.4704 | varas (Texas) per day |
| 42.51960 | varas (Texas) per hour |
| 0.70866 | varas (Texas) per minute |
| 0.011811 | varas (Texas) per second |
| 2,834.6112 | feet per day |
| 118.1088 | feet per hour |
| 1.96848 | feet per minute |
| 0.032808 | feet per second |
| 3,779.4816 | spans per day |
| 157.4784 | spans per hour |
| 2.62464 | spans per minute |
| 0.043744 | spans per second |
| 4,294.8576 | links per day |
| 178.9524 | links per hour |
| 2.98254 | links per minute |
| 0.049709 | links per second |
| 8,503.8336 | hands per day |
| 354.3264 | hands per hour |
| 5.90544 | hands per minute |
| 0.098424 | hands per second |
| 8,640.0 | decimeters per day |
| 360.0 | decimeters per hour |
| 6.0 | decimeters per minute |
| 0.1 | decimeters per second |
| 86,400 | centimeters per day |
| 3,600.0 | centimeters per hour |
| 60.0 | centimeters per minute |
| 1. | centimeters per second |
| 34,015.68 | inches per day |
| 1,417.32 | inches per hour |
| 23.6220 | inches per minute |
| 0.3937 | inches per second |
| 864,000 | millimeters per day |
| 36,000 | millimeters per hour |
| 600 | millimeters per minute |
| 10 | millimeters per second |
| 34,015,680 | mils per day |
| 1,417,320 | mils per hour |
| 23,622 | mils per minute |

## CENTIMETERS PER SECOND (cont'd): =

| | |
|---|---|
| 393.70 | mils per second |
| 864,000,000 | microns per day |
| 36,000,000 | microns per hour |
| 600,000 | microns per minute |
| 10,000 | microns per second |

## CENTIMETER OF MERCURY (O° C): =

| | |
|---|---|
| 0.0013595 | hektometers of water @ 60° F. |
| 0.013595 | dekameters of water @ 60° F. |
| 0.13595 | meters of water @ 60° F. |
| 0.44604 | feet of water @ 60° F. |
| 0.44604 | ounces of water @ 60° F. |
| 5.35248 | inches of water @ 60° F. |
| 1.35952 | decimeters of water @ 60° F. |
| 13.595299 | centimeters of water @ 60° F. |
| 135.95299 | millimeters of water @ 60° F. |
| 0.0001 | hektometers of mercury @ 32° F. |
| 0.001 | dekameters of mercury @ 32° F. |
| 0.01 | meters of mercury @ 32° F. |
| 0.032808 | feet of mercury @ 32° F. |
| 0.032808 | ounces of mercury @ 32° F. |
| 0.3937 | inches of mercury @ 32° F. |
| 0.1 | decimeters of mercury @ 32° F. |
| 1.0 | centimeters of mercury @ 32° F. |
| 10.0 | millimeters of mercury @ 32° F. |
| 1,498.62505 | tons per square hektometer |
| 14.98625 | tons per square dekameter |
| 0.14986 | tons per square meter |
| 0.013923 | tons per square foot |
| 0.000096685 | tons per square inch |
| 0.0014986 | tons per square decimeter |
| 0.000014986 | tons per square centimeter |
| 0.00000014986 | tons per square millimeter |
| 1,359,529.9 | kilograms per square hektometer |
| 13,595.299 | kilograms per square dekameter |
| 135.95299 | kilograms per square meter |
| 12.63034 | kilograms per square foot |
| 0.087711 | kilograms per square inch |
| 1.35953 | kilograms per square decimeter |
| 0.013595 | kilograms per square centimeter |
| 0.00013595 | kilograms per square millimeter |
| 2,997,249.93506 | pounds per square hektometer |
| 29,972.49935 | pounds per square dekameter |

## CENTIMETER OF MERCURY ($0^\circ$ C) (cont'd): =

| | |
|---|---|
| 299.7250 | pounds per square meter |
| 27.845 | pounds per square foot |
| 0.19337 | pounds per square inch |
| 2.99725 | pounds per square decimeter |
| 0.029973 | pounds per square centimeter |
| 0.00029973 | pounds per square millimeter |
| 13,595,299 | hektograms per square hektometer |
| 135,952.99 | hektograms per square dekameter |
| 1,359.5299 | hektograms per square meter |
| 126.3034 | hektograms per square foot |
| 0.87711 | hektograms per square inch |
| 13.5953 | hektograms per square decimeter |
| 0.13595 | hektograms per square centimeter |
| 0.0013595 | hektograms per square millimeter |
| 135,952,990 | dekagrams per square hektometer |
| 1,359,529.9 | dekagrams per square dekameter |
| 13,595.299 | dekagrams per square meter |
| 1,263.034 | dekagrams per square foot |
| 8.7711 | dekagrams per square inch |
| 135.95299 | dekagrams per square decimeter |
| 1.35953 | dekagrams per square centimeter |
| 0.013595 | dekagrams per square millimeter |
| 47,955,999 | ounces per square hektometer |
| 479,560 | ounces per square dekameter |
| 4,795.6 | ounces per square meter |
| 445.520 | ounces per square foot |
| 3.09392 | ounces per square inch |
| 47.9560 | ounces per square decimeter |
| 0.479560 | ounces per square centimeter |
| 0.00479560 | ounces per square millimeter |
| 1,359,529,900 | grams per square hektometer |
| 13,595,299 | grams per square dekameter |
| 135,952.99 | grams per square meter |
| 12,630.34 | grams per square foot |
| 87.71111 | grams per square inch |
| 1,359.5299 | grams per square decimeter |
| 13.595299 | grams per square centimeter |
| 0.13595 | grams per square millimeter |
| 13,595,299,000 | decigrams per square hektometer |
| 135,952,990 | decigrams per square dekameter |
| 1,359,529.9 | decigrams per square meter |
| 126,303.4 | decigrams per square foot |
| 877.11111 | decigrams per square inch |
| 13,595.299 | decigrams per square decimeter |

## CENTIMETER OF MERCURY ($0^\circ$ C) (cont'd): =

| | |
|---|---|
| 135.95299 | decigrams per square centimeter |
| 1.35953 | decigrams per square millimeter |
| 135,952,990,000 | centigrams per square hektometer |
| 1,359,529,900 | centigrams per square dekameter |
| 13,595,299 | centigrams per square meter |
| 1,263,034 | centigrams per square foot |
| 8,771.1111 | centigrams per square inch |
| 135,952.99 | centigrams per square decimeter |
| 1,359.5299 | centigrams per square centimeter |
| 13.595299 | centigrams per square millimeter |
| 1,359,529,900,000 | milligrams per square hektometer |
| 13,595,299,000 | milligrams per square dekameter |
| 135,952,990 | milligrams per square meter |
| 12,630,340 | milligrams per square foot |
| 87,711.1111 | milligrams per square inch |
| 1,359,529.9 | milligrams per square decimeter |
| 13,595.299 | milligrams per square centimeter |
| 135.95299 | milligrams per square millimeter |
| 0.013333 | bars |
| 0.013158 | atmospheres |
| 1,322,220,000,000 | dynes per square hektometer |
| 13,332,200,000 | dynes per square dekameter |
| 133,322,000 | dynes per square meter |
| 12,385,916.01 | dynes per square foot |
| 86,013.3056 | dynes per square inch |
| 1,333,220 | dynes per square decimeter |
| 13,332.20 | dynes per square centimeter |
| 133.3220 | dynes per square millimeter |

## CENTIMETER PER SECOND PER SECOND: =

| | |
|---|---|
| 0.000006214 | miles per second per second |
| 0.00001 | kilometers per second per second |
| 0.000049709 | furlongs per second per second |
| 0.0001 | hektometers per second per second |
| 0.00049709 | chains per second per second |
| 0.001 | dekameters per second per second |
| 0.0019884 | rods per second per second |
| 0.01 | meters per second per second |
| 0.010936 | yards per second per second |
| 0.011811 | varas (Texas) per second per second |
| 0.032808 | feet per second per second |
| 0.043744 | spans per second per second |

## CENTIMETER PER SECOND PER SECOND (cont'd): =

0.049709 . . . . . . . . . . . . . . . . . . . . . . .links per second per second
0.098424 . . . . . . . . . . . . . . . . . . . . . . . hands per second per second
0.10 . . . . . . . . . . . . . . . . . . . . . . . . . decimeters per second per second
0.3937 . . . . . . . . . . . . . . . . . . . . . . . .inches per second per second
1.00 . . . . . . . . . . . . . . . . . . . . . . . .centimeters per second per second
10.00 . . . . . . . . . . . . . . . . . . . . . . . millimeters per second per second
393.70 . . . . . . . . . . . . . . . . . . . . . . . . mils per second per second
10,000 . . . . . . . . . . . . . . . . . . . . . . .microns per second per second

## CHAIN (SURVEYOR'S OR GUNTER'S): =

0.0125 . . . . . . . . . . . . . . . . . . . . . . . . . . . . . . . . . . . . . . .miles
0.020117 . . . . . . . . . . . . . . . . . . . . . . . . . . . . . . . . . . kilometers
0.1 . . . . . . . . . . . . . . . . . . . . . . . . . . . . . . . . . . . . . . . furlongs
0.20117 . . . . . . . . . . . . . . . . . . . . . . . . . . . . . . . . . hektometers
2.0117 . . . . . . . . . . . . . . . . . . . . . . . . . . . . . . . . . . dekameters
4 . . . . . . . . . . . . . . . . . . . . . . . . . . . . . . . . . . . . . . . . . . . rods
20.117 . . . . . . . . . . . . . . . . . . . . . . . . . . . . . . . . . . . . . .meters
22 . . . . . . . . . . . . . . . . . . . . . . . . . . . . . . . . . . . . . . . . . yards
23.76 . . . . . . . . . . . . . . . . . . . . . . . . . . . . . . . . . varas (Texas)
66 . . . . . . . . . . . . . . . . . . . . . . . . . . . . . . . . . . . . . . . . . . feet
88 . . . . . . . . . . . . . . . . . . . . . . . . . . . . . . . . . . . . . . . . . spans
100 . . . . . . . . . . . . . . . . . . . . . . . . . . . . . . . . . . . . . . . . .links
198 . . . . . . . . . . . . . . . . . . . . . . . . . . . . . . . . . . . . . . . . hands
201.17 . . . . . . . . . . . . . . . . . . . . . . . . . . . . . . . . . . . decimeters
2,011.7 . . . . . . . . . . . . . . . . . . . . . . . . . . . . . . . . . .centimeters
792 . . . . . . . . . . . . . . . . . . . . . . . . . . . . . . . . . . . . . . . .inches
20,117 . . . . . . . . . . . . . . . . . . . . . . . . . . . . . . . . . . millimeters
792,000 . . . . . . . . . . . . . . . . . . . . . . . . . . . . . . . . . . . . . . mils
21,117,000 . . . . . . . . . . . . . . . . . . . . . . . . . . . . . . . . . . .microns
21,117,000,000 . . . . . . . . . . . . . . . . . . . . . . . . . . . . milli microns
21,117,000,000 . . . . . . . . . . . . . . . . . . . . . . . . . micro millimeters
31,244,672 . . . . . . . . . . . . . . . . . . . wave lengths of red line of cadmium
$21{,}117 \times 10^{9}$ . . . . . . . . . . . . . . . . . . . . . . . . . . Angstrom Units

## CIRCULAR MIL: =

0.0000000000000000019564 . . . . . . . . . . . . . . .square miles or sections
0.0000000000000000050671 . . . . . . . . . . . . . . . . square kilometers
0.000000000000012521 . . . . . . . . . . . . . . . . . . . . .square furlongs
0.000000000000050671 . . . . . . . . . . . . . . . . . . square hektometers
0.0000000000012521 . . . . . . . . . . . . . . . . . . . . . . . . .square chains

CIRCULAR MIL (cont'd): =

| | |
|---|---|
| 0.000000000050671 | square dekameters |
| 0.000000000020034 | square rods |
| 0.00000000050671 | square meters |
| 0.00000000060602 | square yards |
| 0.00000000070687 | square varas (Texas) |
| 0.0000000054542 | square feet |
| 0.0000000096964 | square spans |
| 0.000000012521 | square links |
| 0.000000049186 | square hands |
| 0.000000050671 | square decimeters |
| 0.0000050671 | square centimeters |
| 0.0000007854 | square inches |
| 0.00050671 | square millimeters |
| 0.0000000000000050671 | hectares |
| 0.0000000000050671 | ares |
| 0.00000000050671 | centares (centiares) |
| 0.7854 | square mils |
| 0.000001 | circular inches |
| 0.001 | inches |
| 0.000645143 | circular millimeters |

CHEVAL-VAPEUR (METRIC HORSEPOWER): =

| | |
|---|---|
| 62,832,926.34 | foot poundals |
| 753,995,116.08 | inch poundals |
| 1,952,910 | foot pounds |
| 23,434,920 | inch pounds |
| 0.69727 | ton (short) calories |
| 632.551 | kilogram calories |
| 1,394.53604 | pound calories |
| 6,325.51 | hektogram calories |
| 63,255.1 | dekagram calories |
| 22,312.57664 | ounce calories |
| 632,551 | gram calories |
| 6,325,510 | decigram calories |
| 63,255,100 | centigram calories |
| 632,551,000 | milligram calories |
| 17.6448 | kilowatt days |
| 0.7352 | kilowatt hours |
| 0.012533 | kilowatt minutes |
| 0.00020422 | kilowatt seconds |
| 30.633333 | watt days |
| 735.2 | watt hours |

CHEVAL-VAPEUR (METRIC HORSEPOWER) (cont'd): =

| | |
|---|---|
| 44,120 | watt minutes |
| 2,646,720 | watt seconds |
| 297.62401 | ton meters |
| 270,000 | kilogram meters |
| 595,248.02 | pound meters |
| 2,700,000 | hektogram meters |
| 27,000,000 | dekagram meters |
| 9,523,968.32 | ounce meters |
| 270,000,000 | gram meters |
| 2,700,000,000 | decigram meters |
| 27,000,000,000 | centigram meters |
| 270,000,000,000 | milligram meters |
| 2.97624 | ton hektometers |
| 2,700 | kilogram hektometers |
| 5,952.4802 | pound hektometers |
| 27,000 | hektogram hektometers |
| 270,000 | dekagram hektometers |
| 95,239.6832 | ounce hektometers |
| 2,700,000 | gram hektometers |
| 27,000,000 | decigram hektometers |
| 270,000,000 | centigram hektometers |
| 2,700,000,000 | milligram hektometers |
| 29.7624 | ton dekameters |
| 27,000 | kilogram dekameters |
| 59,524.802 | pound dekameters |
| 270,000 | hektogram dekameters |
| 2,700,000 | dekagram dekameters |
| 952,396.832 | ounce dekameters |
| 27,000,000 | gram dekameters |
| 270,000,000 | decigram dekameters |
| 2,700,000,000 | centigram dekameters |
| 27,000,000,000 | milligram dekameters |
| 9.07158 | ton feet |
| 82,296 | kilogram feet |
| 181,431.5965 | pound feet |
| 822,960 | hektogram feet |
| 8,229,600 | dekagram feet |
| 2,902,905.54394 | ounce feet |
| 82,296,000 | gram feet |
| 822,960,000 | decigram feet |
| 8,229,600,000 | centigram feet |
| 82,296,000,000 | milligram feet |
| 0.75596 | ton inches |
| 6,858 | kilogram inches |

CHEVAL-VAPEUR (METRIC HORSEPOWER) (cont'd): =

| | |
|---|---|
| 15,119.29971 | pound inches |
| 68,580 | hektogram inches |
| 685,800 | dekagram inches |
| 241,908.79536 | ounce inches |
| 6,858,000 | gram inches |
| 68,580,000 | decigram inches |
| 685,800,000 | centigram inches |
| 6,858,000,000 | milligram inches |
| 2,976.24 | ton decimeters |
| 2,700,000 | kilogram decimeters |
| 5,952,480.2 | pound decimeters |
| 27,000,000 | hektogram decimeters |
| 270,000,000 | dekagram decimeters |
| 95,239,683.2 | ounce decimeters |
| 2,700,000,000 | gram decimeters |
| 27,000,000,000 | decigram decimeters |
| 270,000,000,000 | centigram decimeters |
| 2,700,000,000,000 | milligram decimeters |
| 29,762.4 | ton centimeters |
| 27,000,000 | kilogram centimeters |
| 59,524,802 | pound centimeters |
| 270,000,000 | hektogram centimeters |
| 2,700,000,000 | dekagram centimeters |
| 952,396,832 | ounce centimeters |
| 27,000,000,000 | gram centimeters |
| 270,000,000,000 | decigram centimeters |
| 2,700,000,000,000 | milligram centimeters |
| 297,624 | ton millimeters |
| 270,000,000 | kilogram millimeters |
| 595,248,020 | pound millimeters |
| 2,700,000,000 | hektogram millimeters |
| 27,000,000,000 | dekagram millimeters |
| 9,523,968,320 | ounce millimeters |
| 270,000,000,000 | gram millimeters |
| 2,700,000,000,000 | decigram millimeters |
| 27,000,000,000,000 | milligram millimeters |
| 26.1298 | kiloliter atmospheres |
| 261.298 | hektoliter atmospheres |
| 2,612.98 | dekaliter atmospheres |
| 26,129.8 | liter-atmospheres |
| 261,298 | deciliter-atmospheres |
| 2,612,980 | centiliter-atmospheres |
| 26,129,800 | milliliter-atmospheres |
| 0.0000002613 | cubic kilometer-atmospheres |
| 0.00002613 | cubic hektometer-atmospheres |

CHEVAL-VAPEUR (METRIC HORSPOWER) (cont'd): =

| | |
|---|---|
| 0.02613 | cubic dekameter-atmospheres |
| 26.1298 | cubic meter-atmospheres |
| 922.74776 | cubic feet-atmospheres |
| 26,129.8 | cubic decimeters-atmospheres |
| 26,129,800 | cubic centimeter atmospheres |
| 26,129,800,000 | cubic millimeter atmospheres |
| 2,648,700 | joules |
| 0.0410967 | horsepower days |
| 0.98632 | horsepower hours |
| 59.17920 | horsepower minutes |
| 3,550.752 | horsepower seconds |
| 0.171945 | pounds of carbon oxidized with 100% efficiency |
| 2.5877 | pounds of water evaporated from and at 212° F. |
| 2,510.152 | B.T.U. |

CUBIC CENTIMETER: =

| | |
|---|---|
| 0.00000000000000062137 | cubic miles |
| 0.000000000000001 | cubic kilometers |
| 0.00000000000000122833 | cubic furlongs |
| 0.000000000001 | cubic hektometers |
| 0.00000000122833 | cubic chains |
| 0.000000001 | cubic dekameters |
| 0.0000000078613 | cubic rods |
| 0.000001 | cubic meters |
| 0.000001 | kiloliters |
| 0.0000013079 | cubic yards |
| 0.0000016476 | cubic varas (Texas) |
| 0.0000062897 | barrels |
| 0.00001 | hectoliters |
| 0.000028378 | bushels—U.S. (dry) |
| 0.000027496 | bushels—Imperial (dry) |
| 0.000035314 | cubic feet |
| 0.000083707 | cubic spans |
| 0.0001 | dekaliter |
| 0.000113512 | pecks |
| 0.000122833 | cubic links |
| 0.00026417 | gallons—U.S. (liquid) |
| 0.00022705 | gallons—U.S. (dry) |
| 0.00021997 | gallons—Imperial |
| 0.00095635 | cubic hands |
| 0.0010567 | quarts (liquid) |

**CUBIC CENTIMETER (cont'd): =**

0.00090808 . . . . . . . . . . . . . . . . . . . . . . . . . . . .quarts (dry)
0.001 . . . . . . . . . . . . . . . . . . . . . . . . . . . . . . . . . .liters
0.001 . . . . . . . . . . . . . . . . . . . . . . . . . . . . cubic decimeters
0.0021134 . . . . . . . . . . . . . . . . . . . . . . . . . . . pints (liquid)
0.0018162 . . . . . . . . . . . . . . . . . . . . . . . . . . . . pints (dry)
0.0084536 . . . . . . . . . . . . . . . . . . . . . . . . . . .gills (liquid)
0.01 . . . . . . . . . . . . . . . . . . . . . . . . . . . . . . . deciliters
0.061023 . . . . . . . . . . . . . . . . . . . . . . . . . . . cubic inches
0.1 . . . . . . . . . . . . . . . . . . . . . . . . . . . . . . . .centiliters
1 . . . . . . . . . . . . . . . . . . . . . . . . . . . . . . . . . milliliters
1 . . . . . . . . . . . . . . . . . . . . . . . . . . . . cubic centimeters
1000 . . . . . . . . . . . . . . . . . . . . . . . . . . .cubic millimeters
0.033814 . . . . . . . . . . . . . . . . . . . . . . . . . . ounces (fluid)
0.27051 . . . . . . . . . . . . . . . . . . . . . . . . . . . drams (fluid)

**CUBIC FOOT: =**

0.000000000017596 . . . . . . . . . . . . . . . . . . . . . . . cubic miles
0.000000000028317 . . . . . . . . . . . . . . . . . . . cubic kilometers
0.0000000034783 . . . . . . . . . . . . . . . . . . . . . . cubic furlongs
0.000000028317 . . . . . . . . . . . . . . . . . . . . cubic hektometers
0.0000034783 . . . . . . . . . . . . . . . . . . . . . . . . . cubic chains
0.000028317 . . . . . . . . . . . . . . . . . . . . . . cubic dekameters
0.00022261 . . . . . . . . . . . . . . . . . . . . . . . . . . . .cubic rods
0.028317 . . . . . . . . . . . . . . . . . . . . . . . . . . . cubic meters
0.028317 . . . . . . . . . . . . . . . . . . . . . . . . . . . . . kiloliters
0.037036 . . . . . . . . . . . . . . . . . . . . . . . . . . . . cubic yards
0.046656 . . . . . . . . . . . . . . . . . . . . . . cubic varas (Texas)
0.17811 . . . . . . . . . . . . . . . . . . . . . . . . . . . . . . . .barrels
0.28317 . . . . . . . . . . . . . . . . . . . . . . . . . . . . . . hectoliters
0.80358 . . . . . . . . . . . . . . . . . . . . . . . bushels—U.S. (dry)
0.77860 . . . . . . . . . . . . . . . . . . . . . bushels—Imperial (dry)
2.37033 . . . . . . . . . . . . . . . . . . . . . . . . . . . . . cubic spans
2.8317 . . . . . . . . . . . . . . . . . . . . . . . . . . . . . . . dekaliters
3.2143 . . . . . . . . . . . . . . . . . . . . . . . . . . . . . . . . . . pecks
3.48327 . . . . . . . . . . . . . . . . . . . . . . . . . . . . . . cubic links
7.48050 . . . . . . . . . . . . . . . . . . . . . . gallons—U.S. (liquid)
6.42937 . . . . . . . . . . . . . . . . . . . . . . . . .gallons—U.S. (dry)
6.22889 . . . . . . . . . . . . . . . . . . . . . . . . . .gallons—Imperial
27.08096 . . . . . . . . . . . . . . . . . . . . . . . . . . . . .cubic hands
29.92257 . . . . . . . . . . . . . . . . . . . . . . . . . . . quarts (liquid)
25.71410 . . . . . . . . . . . . . . . . . . . . . . . . . . . . .quarts (dry)
28.317 . . . . . . . . . . . . . . . . . . . . . . . . . . . . . . . . . . .liters
28.317 . . . . . . . . . . . . . . . . . . . . . . . . . . . cubic decimeters

CUBIC FOOT (cont'd): =

| | |
|---|---|
| 59.84515 | pints (liquid) |
| 51.4934 | pints (dry) |
| 239.38060 | gills (liquid) |
| 283.17 | deciliters |
| 1,727.98829 | cubic inches |
| 2,831.7 | centiliters |
| 28,317 | milliliters |
| 28,317 | cubic centimeters |
| 28,317,000 | cubic millimeters |
| 957.51104 | ounces (fluid) |
| 7,660.03167 | drams (fluid) |
| 0.9091 | sacks of cement (set) |
| 62.35 | pounds of water @ 60° F. |
| 64.3 | pounds of salt water |
| 72.0 | pounds of salt water at 60° F. of 1.155 specific gravity |
| 489.542 | pounds of steel of 7.851 specific gravity |

CUBIC FOOT PER DAY: =

| | |
|---|---|
| 0.17811 | barrels per day |
| 0.0074214 | barrels per hour |
| 0.00012369 | barrels per minute |
| 0.0000020615 | barrels per second |
| 0.028317 | kiloliters per day |
| 0.0011799 | kiloliters per hour |
| 0.000019664 | kiloliters per minute |
| 0.00000032774 | kiloliters per second |
| 0.028317 | cubic meters per day |
| 0.0011799 | cubic meters per hour |
| 0.000019664 | cubic meters per minute |
| 0.00000032774 | cubic meters per second |
| 0.037036 | cubic yards per day |
| 0.0015432 | cubic yards per hour |
| 0.00002572 | cubic yards per minute |
| 0.00000042866 | cubic yards per second |
| 0.28317 | hektoliters per day |
| 0.011799 | hektoliters per hour |
| 0.00019664 | hektoliters per minute |
| 0.0000032774 | hektoliters per second |
| 1 | cubic feet per day |
| 0.041666 | cubic feet per hour |

CUBIC FOOT PER DAY (cont'd): =

| | |
|---|---|
| 0.00069444 | cubic feet per minute |
| 0.000011574 | cubic feet per second |
| 2.8317 | dekaliters per day |
| 0.11799 | dekaliters per hour |
| 0.0019664 | dekaliters per minute |
| 0.000032774 | dekaliters per second |
| 7.48050 | gallons per day |
| 0.31169 | gallons per hour |
| 0.0051948 | gallons per minute |
| 0.000086580 | gallons per second |
| 6.22889 | gallons (Imperial) per day |
| 0.25954 | gallons (Imperial) per hour |
| 0.0043256 | gallons (Imperial) per minute |
| 0.000072094 | gallons (Imperial) per second |
| 29.92257 | quarts per day |
| 1.24679 | quarts per hour |
| 0.020780 | quarts per minute |
| 0.00034633 | quarts per second |
| 28.317 | liters per day |
| 1.17988 | liters per hour |
| 0.019664 | liters per minute |
| 0.00032774 | liters per second |
| 28.317 | cubic decimeters per day |
| 1.17988 | cubic decimeters per hour |
| 0.019664 | cubic decimeters per minute |
| 0,00032774 | cubic decimeters per second |
| 59.84515 | pints per day |
| 2.49354 | pints per hour |
| 0.041559 | pints per minute |
| 0.00069265 | pints per second |
| 239.38060 | gills per day |
| 9.97416 | gills per hour |
| 0.166236 | gills per minute |
| 0.0027706 | gills per second |
| 283.17 | deciliters per day |
| 11.79875 | deciliters per hour |
| 0.19664 | deciliters per minute |
| 0.0032774 | deciliters per second |
| 1,727.98829 | cubic inches per day |
| 72.0 | cubic inches per hour |
| 1.20 | cubic inches per minute |
| 0.020 | cubic inches per second |
| 2,831.7 | centiliters per day |
| 117.98750 | centiliters per hour |

## CUBIC FOOT PER DAY (cont'd): =

| | |
|---|---|
| 1.9664 | centiliters per minute |
| 0.032774 | centiliters per second |
| 28,317 | milliliters per day |
| 1,179.8750 | milliliters per hour |
| 19.664 | milliliters per minute |
| 0.32774 | milliliters per second |
| 28,317 | cubic centimeters per day |
| 1,179.8750 | cubic centimeters per hour |
| 19.664 | cubic centimeters per minute |
| 0.32774 | cubic centimeters per second |
| 28,317,000 | cubic millimeters per day |
| 1,179,875 | cubic millimeters per hour |
| 19,664 | cubic millimeters per minute |
| 327.74 | cubic millimeters per second |

## CUBIC FOOT PER HOUR: =

| | |
|---|---|
| 4.27464 | barrels per day |
| 0.17811 | barrels per hour |
| 0.0029685 | barrels per minute |
| 0.000049475 | barrels per second |
| 0.67961 | kiloliters per day |
| 0.028317 | kiloliters per hour |
| 0.00047195 | kiloliters per minute |
| 0.0000078658 | kiloliters per second |
| 0.67961 | cubic meters per day |
| 0.028317 | cubic meters per hour |
| 0.00047195 | cubic meters per minute |
| 0.0000078658 | cubic meters per second |
| 0.88888 | cubic yards per day |
| 0.037036 | cubic yards per hour |
| 0.00061728 | cubic yards per minute |
| 0.000010288 | cubic yards per second |
| 6.79605 | hectoliters per day |
| 0.28317 | hectoliters per hour |
| 0.0047195 | hectoliters per minute |
| 0.000078658 | hectoliters per second |
| 24 | cubic feet per day |
| 1 | cubic feet per hour |
| 0.016667 | cubic feet per minute |
| 0.00027778 | cubic feet per second |
| 67.96051 | dekaliters per day |
| 2.8317 | dekaliters per hour |

**CUBIC FOOT PER HOUR (cont'd): =**

| | |
|---|---|
| 0.047195 | dekaliters per minute |
| 0.00078658 | dekaliters per second |
| 179.53056 | gallons per day |
| 7.48050 | gallons per hour |
| 0.12467 | gallons per minute |
| 0.0020779 | gallons per second |
| 149.48928 | gallons (Imperial) per day |
| 6.22889 | gallons (Imperial) per hour |
| 0.10381 | gallons (Imperial) per minute |
| 0.0017302 | gallons (Imperial) per second |
| 718.13952 | quarts per day |
| 29.92257 | quarts per hour |
| 0.49871 | quarts per minute |
| 0.0083118 | quarts per second |
| 679.60512 | liters per day |
| 28.317 | liters per hour |
| 0.47195 | liters per minute |
| 0.0078658 | liters per second |
| 679.60512 | cubic decimeters per day |
| 28.317 | cubic decimeters per hour |
| 0.47195 | cubic decimeters per minute |
| 0.0078658 | cubic decimeters per second |
| 1,436.31360 | pints per day |
| 59.84515 | pints per hour |
| 0.99744 | pints per minute |
| 0.016624 | pints per second |
| 5,745.25440 | gills per day |
| 239.38060 | gills per hour |
| 3.98976 | gills per minute |
| 0.066496 | gills per second |
| 6,796.05120 | deciliters per day |
| 283.17 | deciliters per hour |
| 4.71948 | deciliters per minute |
| 0.078658 | deciliters per second |
| 41,472.0 | cubic inches per day |
| 1,727.98829 | cubic inches per hour |
| 28.80 | cubic inches per minute |
| 0.480 | cubic inches per second |
| 67,960.512 | centiliters per day |
| 2,831.7 | centiliters per hour |
| 47.19480 | centiliters per minute |
| 0.78658 | centiliters per second |
| 679,605.120 | milliliters per day |
| 28,317 | milliliters per hour |
| 471.94980 | milliliters per minute |

CUBIC FOOT PER HOUR (cont'd): =

7.86583 . . . . . . . . . . . . . . . . . . . . . . . . . . . . . milliliters per second
679,605.120 . . . . . . . . . . . . . . . . . . . . . . . . . .cubic centimeters per day
28,317 . . . . . . . . . . . . . . . . . . . . . . . . . . . cubic centimeters per hour
471.94980 . . . . . . . . . . . . . . . . . . . . . . . .cubic centimeters per minute
7.86583 . . . . . . . . . . . . . . . . . . . . . . . . .cubic centimeters per second
679,605,120 . . . . . . . . . . . . . . . . . . . . . . . . . cubic millimeters per day
28,317,000 . . . . . . . . . . . . . . . . . . . . . . . . cubic millimeters per hour
471,949.80 . . . . . . . . . . . . . . . . . . . . . . . cubic millimeters per minute
7,865.83 . . . . . . . . . . . . . . . . . . . . . . . . cubic millimeters per second

CUBIC FOOT PER MINUTE: =

256.47840 . . . . . . . . . . . . . . . . . . . . . . . . . . . . . . . barrels per day
10.68660 . . . . . . . . . . . . . . . . . . . . . . . . . . . . . . barrels per hour
0.17811 . . . . . . . . . . . . . . . . . . . . . . . . . . . . . barrels per minute
0.0029685 . . . . . . . . . . . . . . . . . . . . . . . . . . . . barrels per second
40.77648 . . . . . . . . . . . . . . . . . . . . . . . . . . . . . . kiloliters per day
1.69902 . . . . . . . . . . . . . . . . . . . . . . . . . . . . . kiloliters per hour
0.028317 . . . . . . . . . . . . . . . . . . . . . . . . . . . kiloliters per minute
0.00047195 . . . . . . . . . . . . . . . . . . . . . . . . . . kiloliters per second
40.77648 . . . . . . . . . . . . . . . . . . . . . . . . . . . cubic meters per day
1.69902 . . . . . . . . . . . . . . . . . . . . . . . . . . . cubic meters per hour
0.028317 . . . . . . . . . . . . . . . . . . . . . . . . . cubic meters per minute
0.00047195 . . . . . . . . . . . . . . . . . . . . . . . .cubic meters per second
53.33213 . . . . . . . . . . . . . . . . . . . . . . . . . . . cubic yards per day
2.22217 . . . . . . . . . . . . . . . . . . . . . . . . . . .cubic yards per hour
0.037036 . . . . . . . . . . . . . . . . . . . . . . . . . cubic yards per minute
0.00061727 . . . . . . . . . . . . . . . . . . . . . . . . cubic yards per second
407.76480 . . . . . . . . . . . . . . . . . . . . . . . . . . . .hectoliters per day
16.99020 . . . . . . . . . . . . . . . . . . . . . . . . . . . hectoliters per hour
0.28317 . . . . . . . . . . . . . . . . . . . . . . . . . . .hectoliters per minute
0.0047195 . . . . . . . . . . . . . . . . . . . . . . . . . .hectoliters per second
1,440 . . . . . . . . . . . . . . . . . . . . . . . . . . . . . . cubic feet per day
60 . . . . . . . . . . . . . . . . . . . . . . . . . . . . . . cubic feet per hour
1 . . . . . . . . . . . . . . . . . . . . . . . . . . . . . cubic feet per minute
0.016667 . . . . . . . . . . . . . . . . . . . . . . . . . . cubic feet per second
4,077.6480 . . . . . . . . . . . . . . . . . . . . . . . . . . . . dekaliters per day
169.9020 . . . . . . . . . . . . . . . . . . . . . . . . . . . .dekaliters per hour
2.8317 . . . . . . . . . . . . . . . . . . . . . . . . . . . dekaliters per minute
0.047195 . . . . . . . . . . . . . . . . . . . . . . . . . . dekaliters per second
10,771.920 . . . . . . . . . . . . . . . . . . . . . . . . . . . . . . gallons per day
448.830 . . . . . . . . . . . . . . . . . . . . . . . . . . . . . . gallons per hour
7.48050 . . . . . . . . . . . . . . . . . . . . . . . . . . . . . gallons per minute

**CUBIC FOOT PER MINUTE (cont'd): =**

| | |
|---|---|
| 0.12468 | gallons per second |
| 8,969.184 | gallons (Imperial) per day |
| 373.7160 | gallons (Imperial) per hour |
| 6.22889 | gallons (Imperial) per minute |
| 0.10381 | gallons (Imperial) per second |
| 43,088.5440 | quarts per day |
| 1,795.3560 | quarts per hour |
| 29.92257 | quarts per minute |
| 0.49871 | quarts per second |
| 40,776.480 | liters per day |
| 1,699.020 | liters per hour |
| 28.3170 | liters per minute |
| 0.47195 | liters per second |
| 40,776.480 | cubic decimeters per day |
| 1,699.020 | cubic decimeters per hour |
| 28.3170 | cubic decimeters per minute |
| 0.47195 | cubic decimeters per second |
| 86,177.0880 | pints per day |
| 3,590.712 | pints per hour |
| 59.84515 | pints per minute |
| 0.99742 | pints per second |
| 344,708.3520 | gills per day |
| 14,362.848 | gills per hour |
| 239.38060 | gills per minute |
| 3.98968 | gills per second |
| 407,764.80 | deciliters per day |
| 16,990.20 | deciliters per hour |
| 283.17 | deciliters per minute |
| 4.7195 | deciliters per second |
| 2,488,303.584 | cubic inches per day |
| 103,769.3160 | cubic inches per hour |
| 1,727.98829 | cubic inches per minute |
| 28.79981 | cubic inches per second |
| 4,077,648.0 | centiliters per day |
| 169,902.0 | centiliters per hour |
| 2,831.7 | centiliters per minute |
| 47.195 | centiliters per second |
| 40,776,480.0 | milliliters per day |
| 1,699,020.0 | milliliters per hour |
| 28,317 | milliliters per minute |
| 471.95 | milliliters per second |
| 40,776,480.0 | cubic centimeters per day |
| 1,699,020.0 | cubic centimeters per hour |
| 28,317 | cubic centimeters per minute |

**CUBIC FOOT PER MINUTE (cont'd): =**

| | |
|---|---|
| 471.95 | cubic centimeters per second |
| 40,776,480,000 | cubic millimeters per day |
| 1,699,020,000 | cubic millimeters per hour |
| 28,317,000 | cubic millimeters per minute |
| 471,950 | cubic millimeters per second |

**CUBIC FOOT PER SECOND: =**

| | |
|---|---|
| 15,388.70400 | barrels per day |
| 641.19600 | barrels per hour |
| 10.68660 | barrels per minute |
| 0.17811 | barrels per second |
| 2,446.58880 | kiloliters per day |
| 101.94120 | kiloliters per hour |
| 1.69902 | kiloliters per minute |
| 0.028317 | kiloliters per second |
| 2,446.5880 | cubic meters per day |
| 101.94120 | cubic meters per hour |
| 1.69902 | cubic meters per minute |
| 0.028317 | cubic meters per second |
| 3,199.91040 | cubic yards per day |
| 133.32960 | cubic yards per hour |
| 2.22216 | cubic yards per minute |
| 0.037036 | cubic yards per second |
| 24,465.8880 | hectoliters per day |
| 1,019.41200 | hectoliters per hour |
| 16.99020 | hectoliters per minute |
| 0.28317 | hectoliters per second |
| 86,400 | cubic feet per day |
| 3,600 | cubic feet per hour |
| 60 | cubic feet per minute |
| 1 | cubic feet per second |
| 244,658.880 | dekaliters per day |
| 10,194.120 | dekaliters per hour |
| 169.9020 | dekaliters per minute |
| 2.8317 | dekaliters per second |
| 646,315.20 | gallons per day |
| 26,929.80 | gallons per hour |
| 448.830 | gallons per minute |
| 7.48050 | gallons per second |
| 538,176.096 | gallons (Imperial) per day |
| 22,424.004 | gallons (Imperial) per hour |
| 373.73340 | gallons (Imperial) per minute |

**CUBIC FOOT PER SECOND (cont'd): =**

| | |
|---|---|
| 6.22889 | gallons (Imperial) per second |
| 2,585,310.0480 | quarts per day |
| 107,721.2520 | quarts per hour |
| 1,795.35420 | quarts per minute |
| 29.92257 | quarts per second |
| 2,446,588.80 | liters per day |
| 101.941.20 | liters per hour |
| 1,699.020 | liters per minute |
| 28.317 | liters per second |
| 2,446,588.80 | cubic decimeters per day |
| 101,941.20 | cubic decimeters per hour |
| 1,699.020 | cubic decimeters per minute |
| 28.317 | cubic decimeters per second |
| 5,170,620.960 | pints per day |
| 215,442.540 | pints per hour |
| 3,590.7090 | pints per minute |
| 59.84515 | pints per second |
| 20,682,483.84 | gills per day |
| 861.770.16 | gills per hour |
| 14,362.8360 | gills per minute |
| 239.38060 | gills per second |
| 24,465,880 | deciliters per day |
| 1,019,412.0 | deciliters per hour |
| 16,990.20 | deciliters per minute |
| 283.17 | deciliters per second |
| 149,298,188.2560 | cubic inches per day |
| 6,220,757.8440 | cubic inches per hour |
| 103,679.29740 | cubic inches per minute |
| 1,727.98829 | cubic inches per second |
| 244,658,880.0 | centiliters per day |
| 10,194,120.0 | centiliters per hour |
| 169,902.0 | centiliters per minute |
| 2,831.7 | centiliters per second |
| 2,446,588,800.0 | milliliters per day |
| 101,941,200.0 | milliliters per hour |
| 1,699,020.0 | milliliters per minute |
| 28,317 | milliliters per second |
| 2,446,588,800.0 | cubic centimeters per day |
| 101,941,200.0 | cubic centimeters per hour |
| 1,699,020.0 | cubic centimeters per minute |
| 28,317 | cubic centimeters per second |
| 2,446,588,800,000 | cubic millimeters per day |
| 101,941,200,000 | cubic millimeters per hour |
| 1,699,020,000 | cubic millimeters per minute |
| 28,317,000 | cubic millimeters per second |

CUBIC INCH: =

| | |
|---|---|
| 0.000000000000010183 | cubic miles |
| 0.000000000000016387 | cubic kilometers |
| 0.0000000000020129 | cubic furlongs |
| 0.000000000016387 | cubic hektometers |
| 0.0000000020129 | cubic chains |
| 0.000000016387 | cubic dekameters |
| 0.00000012883 | cubic rods |
| 0.000016387 | cubic meters |
| 0.000016387 | kiloliters |
| 0.000021434 | cubic yards |
| 0.000027000 | cubic varas (Texas) |
| 0.00010307 | barrels |
| 0.00016387 | hectoliters |
| 0.00046503 | bushels—U.S. (Dry) |
| 0.00045058 | bushels—Imperial (Dry) |
| 0.0005787 | cubic feet |
| 0.0013717 | cubic spans |
| 0.0016387 | dekaliters |
| 0.0018601 | pecks |
| 0.0020129 | cubic links |
| 0.0043290 | gallons—U.S. (Liquid) |
| 0.003721 | gallons—U.S. (Dry) |
| 0.003607 | gallons) Imperial |
| 0.015672 | cubic hands |
| 0.017316 | quarts (Liquid) |
| 0.014881 | quarts (Dry) |
| 0.016387 | liters |
| 0.016387 | cubic decimeters |
| 0.034632 | pints (Liquid) |
| 0.029762 | pints (Dry) |
| 0.13853 | gills (Liquid) |
| 0.16387 | deciliters |
| 1.63871 | centiliters |
| 16.38716 | milliliters |
| 16.38716 | cubic centimeters |
| 16,387.16 | cubic millimeters |
| 4.4329 | drams (fluid) |
| 0.2833 | pounds of steel (specific gravity—7.851) |
| 0.03607 | pounds of water @ 60° F. |
| 0.5541 | ounces of fluid |
| 0.041667 | pounds of salt water @ 60° F. and 1.155 specific gravity |
| 0.0005261 | sacks of cement (set) |

CUBIC INCH PER DAY: =

| | |
|---|---|
| 0.000016387 | kiloliters per day |
| 0.00000068278 | kiloliters per hour |
| 0.000000011380 | kiloliters per minute |
| 0.00000000018966 | kiloliters per second |
| 0.000016387 | cubic meters per day |
| 0.00000068278 | cubic meters per hour |
| 0.000000011380 | cubic meters per minute |
| 0.00000000018966 | cubic meters per second |
| 0.000021434 | cubic yards per day |
| 0.00000089309 | cubic yards per hour |
| 0.000000014885 | cubic yards per minute |
| 0.00000000024808 | cubic yards per second |
| 0.00010307 | barrels per day |
| 0.0000042944 | barrels per hour |
| 0.000000071574 | barrels per minute |
| 0.0000000011929 | barrels per second |
| 0.00016387 | hectoliters per day |
| 0.0000068278 | hectoliters per hour |
| 0.00000011380 | hectoliters per minute |
| 0.0000000018966 | hectoliters per second |
| 0.0005787 | cubic feet per day |
| 0.000024112 | cubic feet per hour |
| 0.00000040187 | cubic feet per minute |
| 0.0000000066979 | cubic feet per second |
| 0.0016387 | dekaliters per day |
| 0.000068278 | dekaliters per hour |
| 0.0000011380 | dekaliters per minute |
| 0.000000018966 | dekaliters per second |
| 0.0043290 | gallons per day |
| 0.00018037 | gallons per hour |
| 0.0000030062 | gallons per minute |
| 0.000000050104 | gallons per second |
| 0.003607 | gallons (Imperial) per day |
| 0.00015029 | gallons (Imperial) per hour |
| 0.0000025049 | gallons (Imperial) per minute |
| 0.000000041748 | gallons (Imperial) per second |
| 0.016387 | liters per day |
| 0.00068278 | liters per hour |
| 0.000011380 | liters per minute |
| 0.00000018966 | liters per second |
| 0.016387 | cubic decimeters per day |
| 0.00068278 | cubic decimeters per hour |
| 0.000011380 | cubic decimeters per minute |
| 0.00000018966 | cubic decimeters per second |

CUBIC INCH PER DAY (cont'd): =

| | |
|---|---|
| 0.017316 | quarts per day |
| 0.00072151 | quarts per hour |
| 0.000012025 | quarts per minute |
| 0.00000020042 | quarts per second |
| 0.034632 | pints per day |
| 0.0014430 | pints per hour |
| 0.000024050 | pints per minute |
| 0.00000040083 | pints per second |
| 0.13853 | gills per day |
| 0.0057722 | gills per hour |
| 0.000096024 | gills per minute |
| 0.0000016034 | gills per second |
| 0.16387 | deciliters per day |
| 0.0068278 | deciliters per hour |
| 0.00011380 | deciliters per minute |
| 0.0000018966 | deciliters per second |
| 1.0 | cubic inches per day |
| 0.041666 | cubic inches per hour |
| 0.00069444 | cubic inches per minute |
| 0.000011574 | cubic inches per second |
| 1.63871 | centiliters per day |
| 0.068278 | centiliters per hour |
| 0.0011380 | centiliters per minute |
| 0.000018966 | centiliters per second |
| 16.38716 | milliliters per day |
| 0.68278 | milliliters per hour |
| 0.011380 | milliliters per minute |
| 0.00018966 | milliliters per second |
| 16.38716 | cubic centimeters per day |
| 0.68278 | cubic centimeters per hour |
| 0.011380 | cubic centimeters per minute |
| 0.00018966 | cubic centimeters per second |
| 16,387.16 | cubic millimeters per day |
| 682.7760 | cubic millimeters per hour |
| 11.3796 | cubic millimeters per minute |
| 0.18966 | cubic millimeters per second |

CUBIC INCH PER HOUR: =

| | |
|---|---|
| 0.00039328 | kiloliters per day |
| 0.000016387 | kiloliters per hour |
| 0.00000027311 | kiloliters per minute |
| 0.0000000045519 | kiloliters per second |

CUBIC INCH PER HOUR (cont'd): =

| | |
|---|---|
| 0.00039328 | cubic meters per day |
| 0.000016387 | cubic meters per hour |
| 0.00000027311 | cubic meters per minute |
| 0.0000000045519 | cubic meters per second |
| 0.00051442 | cubic yards per day |
| 0.000021434 | cubic yards per hour |
| 0.00000035723 | cubic yards per minute |
| 0.0000000059539 | cubic yards per second |
| 0.0024737 | barrels per day |
| 0.00010307 | barrels per hour |
| 0.0000017179 | barrels per minute |
| 0.000000028631 | barrels per second |
| 0.0039328 | hectoliters per day |
| 0.00016387 | hectoliters per hour |
| 0.0000027311 | hectoliters per minute |
| 0.000000045519 | hectoliters per second |
| 0.013889 | cubic feet per day |
| 0.0005787 | cubic feet per hour |
| 0.000009645 | cubic feet per minute |
| 0.00000016075 | cubic feet per second |
| 0.039328 | dekaliters per day |
| 0.0016387 | dekaliters per hour |
| 0.000027311 | dekaliters per minute |
| 0.00000045519 | dekaliters per second |
| 0.10390 | gallons per day |
| 0.0043290 | gallons per hour |
| 0.000072150 | gallons per minute |
| 0.0000012025 | gallons per second |
| 0.086564 | gallons (Imperial) per day |
| 0.003607 | gallons (Imperial) per hour |
| 0.000060114 | gallons (Imperial) per minute |
| 0.0000010019 | gallons (Imperial) per second |
| 0.39328 | liters per day |
| 0.016387 | liters per hour |
| 0.00027311 | liters per minute |
| 0.0000045519 | liters per second |
| 0.39328 | cubic decimeters per day |
| 0.016387 | cubic decimeters per hour |
| 0.00027311 | cubic decimeters per minute |
| 0.0000045519 | cubic decimeters per second |
| 0.41558 | quarts per day |
| 0.017316 | quarts per hour |
| 0.0002886 | quarts per minute |
| 0.000004810 | quarts per second |

CUBIC INCH PER HOUR (cont'd): =

| | |
|---|---|
| 0.83117 | pints per day |
| 0.034632 | pints per hour |
| 0.0005772 | pints per minute |
| 0.000009620 | pints per second |
| 3.32476 | gills per day |
| 0.13853 | gills per hour |
| 0.0023089 | gills per minute |
| 0.000038481 | gills per second |
| 3.93284 | deciliters per day |
| 0.16387 | deciliters per hour |
| 0.0027311 | deciliters per minute |
| 0.000045519 | deciliters per second |
| 24 | cubic inches per day |
| 1.0 | cubic inches per hour |
| 0.016667 | cubic inches per minute |
| 0.00027778 | cubic inches per second |
| 39.32842 | centiliters per day |
| 1.63871 | centiliters per hour |
| 0.027311 | centiliters per minute |
| 0.00045519 | centiliters per second |
| 393.28416 | milliliters per day |
| 16.38716 | milliliters per hour |
| 0.27311 | milliliters per minute |
| 0.0045519 | milliliters per second |
| 393.28416 | cubic centimeters per day |
| 16.38716 | cubic centimeters per hour |
| 0.27311 | cubic centimeters per minute |
| 0.0045519 | cubic centimeters per second |
| 393,284.16 | cubic millimeters per day |
| 16,387.16 | cubic millimeters per hour |
| 273.114 | cubic millimeters per minute |
| 4.5519 | cubic millimeters per second |

CUBIC INCH PER MINUTE: =

| | |
|---|---|
| 0.023598 | kiloliters per day |
| 0.00098323 | kiloliters per hour |
| 0.000016387 | kiloliters per minute |
| 0.00000027312 | kiloliters per second |
| 0.023598 | cubic meters per day |
| 0.00098323 | cubic meters per hour |
| 0.000016387 | cubic meters per minute |
| 0.00000027312 | cubic meters per second |

CUBIC INCH PER MINUTE (cont'd ): =

| | |
|---|---|
| 0.030865 | cubic yards per day |
| 0.0012860 | cubic yards per hour |
| 0.000021434 | cubic yards per minute |
| 0.00000035723 | cubic yards per second |
| 0.14842 | barrels per day |
| 0.0061841 | barrels per hour |
| 0.00010307 | barrels per minute |
| 0.0000017178 | barrels per second |
| 0.23598 | hectoliters per day |
| 0.0098323 | hectoliters per hour |
| 0.00016387 | hectoliters per minute |
| 0.0000027312 | hectoliters per second |
| 0.83333 | cubic feet per day |
| 0.034722 | cubic feet per hour |
| 0.0005787 | cubic feet per minute |
| 0.0000096450 | cubic feet per second |
| 2.35976 | dekaliters per day |
| 0.098323 | dekaliters per hour |
| 0.0016387 | dekaliters per minute |
| 0.000027312 | dekaliters per second |
| 6.23376 | gallons per day |
| 0.25974 | gallons per hour |
| 0.0043290 | gallons per minute |
| 0.000072150 | gallons per second |
| 5.19411 | gallons (Imperial) per day |
| 0.21642 | gallons (Imperial) per hour |
| 0.003607 | gallons (Imperial) per minute |
| 0.000060117 | gallons (Imperial) per second |
| 23.59757 | liters per day |
| 0.98323 | liters per hour |
| 0.016387 | liters per minute |
| 0.00027312 | liters per second |
| 23.59757 | cubic decimeters per day |
| 0.98323 | cubic decimeters per hour |
| 0.016387 | cubic decimeters per minute |
| 0.00027312 | cubic decimeters per second |
| 24.93504 | quarts per day |
| 1.038960 | quarts per hour |
| 0.017316 | quarts per minute |
| 0.0002886 | quarts per second |
| 49.87008 | pints per day |
| 2.07792 | pints per hour |
| 0.034632 | pints per minute |
| 0.0005772 | pints per second |

**CUBIC INCH PER MINUTE (cont'd): =**

| | |
|---|---|
| 199.48032 | gills per day |
| 8.31168 | gills per hour |
| 0.13853 | gills per minute |
| 0.0023088 | gills per second |
| 235.97568 | deciliters per day |
| 9.83232 | deciliters per hour |
| 0.16387 | deciliters per minute |
| 0.0027312 | deciliters per second |
| 1,440 | cubic inches per day |
| 60 | cubic inches per hour |
| 1 | cubic inch per minute |
| 0.016667 | cubic inches per second |
| 2,359.7568 | centiliters per day |
| 98.3232 | centiliters per hour |
| 1.63871 | centiliters per minute |
| 0.027312 | centiliters per second |
| 23,597.568 | milliliters per day |
| 983.232 | milliliters per hour |
| 16.38716 | milliliters per minute |
| 0.27312 | milliliters per second |
| 23,597.568 | cubic centimeters per day |
| 983.232 | cubic centimeters per hour |
| 16.38716 | cubic centimeters per minute |
| 0.27312 | cubic centimeters per second |
| 23,597,568 | cubic millimeters per day |
| 983,232 | cubic millimeters per hour |
| 16,387.16 | cubic millimeters per minute |
| 273.11933 | cubic millimeters per second |

**CUBIC INCH PER SECOND: =**

| | |
|---|---|
| 1.41584 | kiloliters per day |
| 0.058993 | kiloliters per hour |
| 0.00098322 | kiloliters per minute |
| 0.000016387 | kiloliters per second |
| 1.41584 | cubic meters per day |
| 0.058993 | cubic meters per hour |
| 0.00098322 | cubic meters per minute |
| 0.000016387 | cubic meters per second |
| 1.85190 | cubic yards per day |
| 0.077162 | cubic yards per hour |
| 0.0012860 | cubic yards per minute |
| 0.000021434 | cubic yards per second |

CUBIC INCH PER SECOND (cont'd): =

| | |
|---|---|
| 8.90525 | barrels per day |
| 0.37105 | barrels per hour |
| 0.0061842 | barrels per minute |
| 0.00010307 | barrels per second |
| 14.15837 | hectoliters per day |
| 0.58993 | hectoliters per hour |
| 0.0098322 | hectoliters per minute |
| 0.00016387 | hectoliters per second |
| 49.99968 | cubic feet per day |
| 2.08332 | cubic feet per hour |
| 0.034722 | cubic feet per minute |
| 0.0005787 | cubic feet per second |
| 141.58368 | dekaliters per day |
| 5.89932 | dekaliters per hour |
| 0.098332 | dekaliters per minute |
| 0.0016387 | dekaliters per second |
| 374.02560 | gallons per day |
| 15.58440 | gallons per hour |
| 0.25974 | gallons per minute |
| 0.0043290 | gallons per second |
| 311.64480 | gallons (Imperial) per day |
| 12.98520 | gallons (Imperial) per hour |
| 0.21642 | gallons (Imperial) per minute |
| 0.003607 | gallons (Imperial) per second |
| 1,415.83680 | liters per day |
| 58.99320 | liters per hour |
| 0.98322 | liters per minute |
| 0.016387 | liters per second |
| 1,415.83680 | cubic decimeters per day |
| 58.99320 | cubic decimeters per hour |
| 0.98322 | cubic decimeters per minute |
| 0.016387 | cubic decimeters per second |
| 1,496.10240 | quarts per day |
| 62.33760 | quarts per hour |
| 1.038960 | quarts per minute |
| 0.017316 | quarts per second |
| 2,992.20480 | pints per day |
| 124.6752 | pints per hour |
| 2.077920 | pints per minute |
| 0.034632 | pints per second |
| 11,968.9920 | gills per day |
| 498.70800 | gills per hour |
| 8.31180 | gills per minute |
| 0.13853 | gills per second |

CUBIC INCH PER SECOND (cont'd): =

14,158.36800 . . . . . . . . . . . . . . . . . . . . . . . . . . .deciliters per day
589.93200 . . . . . . . . . . . . . . . . . . . . . . . . . . . . deciliters per hour
9.83326 . . . . . . . . . . . . . . . . . . . . . . . . . . . . .deciliters per minute
0.16387 . . . . . . . . . . . . . . . . . . . . . . . . . . . . .deciliters per second
86,400 . . . . . . . . . . . . . . . . . . . . . . . . . . . . .cubic inches per day
3,600 . . . . . . . . . . . . . . . . . . . . . . . . . . . . . cubic inches per hour
60 . . . . . . . . . . . . . . . . . . . . . . . . . . . . . .cubic inches per minute
1.0 . . . . . . . . . . . . . . . . . . . . . . . . . . . . . .cubic inches per second
141,584.5440 . . . . . . . . . . . . . . . . . . . . . . . . . . . centiliters per day
5,899.3560 . . . . . . . . . . . . . . . . . . . . . . . . . . . centiliters per hour
98.32260 . . . . . . . . . . . . . . . . . . . . . . . . . . . centiliters per minute
1.63871 . . . . . . . . . . . . . . . . . . . . . . . . . . . . centiliters per second
1,415,850.6240 . . . . . . . . . . . . . . . . . . . . . . . . . . milliliters per day
58,993.7760 . . . . . . . . . . . . . . . . . . . . . . . . . . .milliliters per hour
983.22960 . . . . . . . . . . . . . . . . . . . . . . . . . . milliliters per minute
16.38716 . . . . . . . . . . . . . . . . . . . . . . . . . . . milliliters per second
1,415,850.6240 . . . . . . . . . . . . . . . . . . . . . . .cubic centimeters per day
58,993.7760 . . . . . . . . . . . . . . . . . . . . . . . cubic centimeters per hour
983.22960 . . . . . . . . . . . . . . . . . . . . . . .cubic centimeters per minute
16.38716 . . . . . . . . . . . . . . . . . . . . . . . .cubic centimeters per second
1,415,850,624 . . . . . . . . . . . . . . . . . . . . . . . .cubic millimeters per day
58,993,776 . . . . . . . . . . . . . . . . . . . . . . . . cubic millimeters per hour
983,229.6 . . . . . . . . . . . . . . . . . . . . . . . .cubic millimeters per minute
16,387.16 . . . . . . . . . . . . . . . . . . . . . . . . cubic millimeters per second

CUBIC METER: =

0.00000000062139 . . . . . . . . . . . . . . . . . . . . . . . . . . . . cubic miles
0.000000001 . . . . . . . . . . . . . . . . . . . . . . . . . . . . cubic kilometers
0.00000012283 . . . . . . . . . . . . . . . . . . . . . . . . . . . . cubic furlongs
0.000001 . . . . . . . . . . . . . . . . . . . . . . . . . . . . .cubic hektometers
0.00012283 . . . . . . . . . . . . . . . . . . . . . . . . . . . . . . cubic chains
0.001 . . . . . . . . . . . . . . . . . . . . . . . . . . . . . . cubic dekameters
0.0078613 . . . . . . . . . . . . . . . . . . . . . . . . . . . . . . . .cubic rods
1 . . . . . . . . . . . . . . . . . . . . . . . . . . . . . . . . . . . . kiloliters
1.307943 . . . . . . . . . . . . . . . . . . . . . . . . . . . . . . . . cubic yards
1.64763 . . . . . . . . . . . . . . . . . . . . . . . . . . . . cubic varas (Texas)
6.28994 . . . . . . . . . . . . . . . . . . . . . . . . . . . . . . . . . . . .barrels
10 . . . . . . . . . . . . . . . . . . . . . . . . . . . . . . . . . . . hectoliters
28.37798 . . . . . . . . . . . . . . . . . . . . . . . . . . . . bushels (U.S.) dry
27.49582 . . . . . . . . . . . . . . . . . . . . . . . . . .bushels (Imperial) dry
35.314445 . . . . . . . . . . . . . . . . . . . . . . . . . . . . . . . . . cubic feet
83.70688 . . . . . . . . . . . . . . . . . . . . . . . . . . . . . . . . . cubic spans

CUBIC METER (cont'd): =

| | |
|---|---|
| 100 | dekaliters |
| 113.51120 | pecks |
| 122.83316 | cubic links |
| 264.17762 | gallons (U.S.) liquid |
| 227.026407 | gallons (U.S.) dry |
| 219.97542 | gallons (Imperial) |
| 956.34894 | cubic hands |
| 1,000 | liters |
| 1,000 | cubic decimeters |
| 1,056.71088 | quarts (liquid) |
| 908.10299 | quarts (dry) |
| 2,113.42176 | pints (liquid) |
| 1,816.19834 | pints (dry) |
| 8,453.68704 | gills (liquid) |
| 10,000 | deciliters |
| 61,022.93879 | cubic inches |
| 100,000 | centiliters |
| 1,000,000 | milliliters |
| 1,000,000 | cubic centimeters |
| 1,000,000,000 | cubic millimeters |
| 2,204.62 | pounds of water @ 39° F. |
| 2,201.82790 | pounds of water @ 60° F. |
| 2,542.608 | pounds of salt water @ 60° F. and 1.155 specific gravity |
| 32.10396 | sacks of cement (set) |
| 33,813.54487 | ounces (fluid) |
| 270,506.35839 | drams (fluid) |

CUBIC METERS PER DAY: =

| | |
|---|---|
| 1 | kiloliters per day |
| 0.041667 | kiloliters per hour |
| 0.00069444 | kiloliters per minute |
| 0.000011574 | kiloliters per second |
| 1 | cubic meters per day |
| 0.041667 | cubic meters per hour |
| 0.00069444 | cubic meters per minute |
| 0.000011574 | cubic meters per second |
| 1.30794 | cubic yards per day |
| 0.054497 | cubic yards per hour |
| 0.00090828 | cubic yards per minute |
| 0.000015138 | cubic yards per second |
| 6.289943 | barrels per day |

CUBIC METERS PER DAY (cont'd): =

| | |
|---|---|
| 0.26208 | barrels per hour |
| 0.0043680 | barrels per minute |
| 0.000072800 | barrels per second |
| 10 | hectoliters per day |
| 0.41667 | hectoliters per hour |
| 0.0069444 | hectoliters per minute |
| 0.00011574 | hectoliters per second |
| 35.31444 | cubic feet per day |
| 1.47143 | cubic feet per hour |
| 0.024524 | cubic feet per minute |
| 0.00040873 | cubic feet per second |
| 100 | dekaliters per day |
| 4.1666 | dekaliters per hour |
| 0.069444 | dekaliters per minute |
| 0.0011574 | dekaliters per second |
| 264.17762 | gallons per day |
| 11.0074008 | gallons per hour |
| 0.18346 | gallons per minute |
| 0.0030576 | gallons per second |
| 219.97542 | gallons (Imperial) per day |
| 9.16564 | gallons (Imperial) per hour |
| 0.15276 | gallons (Imperial) per minute |
| 0.0025460 | gallons (Imperial) per second |
| 1,000 | liters per day |
| 41.66640 | liters per hour |
| 0.69444 | liters per minute |
| 0.011574 | liters per second |
| 1,000 | cubic decimeters per day |
| 41.66640 | cubic decimeters per hour |
| 0.69444 | cubic decimeters per minute |
| 0.011574 | cubic decimeters per second |
| 1,056.71088 | quarts per day |
| 44.02962 | quarts per hour |
| 0.73383 | quarts per minute |
| 0.012230 | quarts per second |
| 2,113.42176 | pints per day |
| 88.059240 | pints per hour |
| 1.46765 | pints per minute |
| 0.024461 | pints per second |
| 8,453.68704 | gills per day |
| 352.23696 | gills per hour |
| 5.87062 | gills per minute |
| 0.097844 | gills per second |
| 10,000 | deciliters per day |

## CUBIC METERS PER DAY (cont'd): =

416.66400 . . . . . . . . . deciliters per hour
6.94440 . . . . . . . . . deciliters per minute
0.11574 . . . . . . . . . deciliters per second
61,022.93879 . . . . . . . . . cubic inches per day
2,542.6080 . . . . . . . . . cubic inches per hour
42.3768 . . . . . . . . . cubic inches per minute
0.70628 . . . . . . . . . cubic inches per second
100,000 . . . . . . . . . centiliters per day
4,166.6400 . . . . . . . . . centiliters per hour
69.4440 . . . . . . . . . centiliters per minute
1.15740 . . . . . . . . . centiliters per second
1,000,000 . . . . . . . . . milliliters per day
41,666.40 . . . . . . . . . milliliters per hour
694.440 . . . . . . . . . milliliters per minute
11.57407 . . . . . . . . . milliliters per second
1,000,000 . . . . . . . . . cubic centimeters per day
4,166.64000 . . . . . . . . . cubic centimeters per hour
69.4440 . . . . . . . . . cubic centimeters per minute
1.15740 . . . . . . . . . cubic centimeters per second
1,000,000,000 . . . . . . . . . cubic millimeters per day
4,166,640 . . . . . . . . . cubic millimeters per hour
69,444 . . . . . . . . . cubic millimeters per minute
1,157.4 . . . . . . . . . cubic millimeters per second

## CUBIC METERS PER HOUR: =

24 . . . . . . . . . kiloliters per day
1 . . . . . . . . . kiloliters per hour
0.016667 . . . . . . . . . kiloliters per minute
0.00027778 . . . . . . . . . kiloliters per second
24 . . . . . . . . . cubic meters per day
1 . . . . . . . . . cubic meters per hour
0.01667 . . . . . . . . . cubic meters per minute
0.00027778 . . . . . . . . . cubic meters per second
31.39085 . . . . . . . . . cubic yards per day
1.30794 . . . . . . . . . cubic yards per hour
0.021799 . . . . . . . . . cubic yards per minute
0.00036332 . . . . . . . . . cubic yards per second
150.95863 . . . . . . . . . barrels per day
6.289943 . . . . . . . . . barrels per hour
0.10483 . . . . . . . . . barrels per minute
0.0017472 . . . . . . . . . barrels per second
240 . . . . . . . . . hectoliters per day

CUBIC METERS PER HOUR (cont'd): =

| | |
|---|---|
| 10 | hectoliters per hour |
| 0.16667 | hectoliters per minute |
| 0.0027778 | hectoliters per second |
| 847.54944 | cubic feet per day |
| 35.31444 | cubic feet per hour |
| 0.58858 | cubic feet per minute |
| 0.0098096 | cubic feet per second |
| 2,400 | dekaliters per day |
| 100 | dekaliters per hour |
| 1.66668 | dekaliters per minute |
| 0.027778 | dekaliters per second |
| 6,340.26288 | gallons per day |
| 264.17762 | gallons per hour |
| 4.40296 | gallons per minute |
| 0.073383 | gallons per second |
| 5,279.41008 | gallons (Imperial) per day |
| 219.97542 | gallons (Imperial) per hour |
| 3.66626 | gallons (Imperial) per minute |
| 0.06114 | gallons (Imperial) per second |
| 24,000 | liters per day |
| 1,000 | liters per hour |
| 16.66680 | liters per minute |
| 0.27778 | liters per second |
| 24,000 | cubic decimeters per day |
| 1,000 | cubic decimeters per hour |
| 16.66680 | cubic decimeters per minute |
| 0.27778 | cubic decimeters per second |
| 25,360.9920 | quarts per day |
| 1,056.71088 | quarts per hour |
| 17.61180 | quarts per minute |
| 0.29353 | quarts per second |
| 50,722.12224 | pints per day |
| 2,113.42176 | pints per hour |
| 35.22370 | pints per minute |
| 0.58706 | pints per second |
| 202,888.48896 | gills per day |
| 8,453.68704 | gills per hour |
| 140.89478 | gills per minute |
| 2.34825 | gills per second |
| 240,000 | deciliters per day |
| 10,000 | deciliters per hour |
| 166.6680 | deciliters per minute |
| 2.77778 | deciliters per second |
| 1,464,550.8480 | cubic inches per day |

## CUBIC METERS PER HOUR (cont'd): =

| | |
|---|---|
| 61,022.93879 | cubic inches per hour |
| 1,017.04920 | cubic inches per minute |
| 16.95082 | cubic inches per second |
| 2,400,000 | centiliters per day |
| 100,000 | centiliters per hour |
| 1,666.680 | centiliters per minute |
| 27.77778 | centiliters per second |
| 24,000,000 | milliliters per day |
| 1,000,000 | milliliters per hour |
| 16,666.80 | milliliters per minute |
| 277.77778 | milliliters per second |
| 24,000,000 | cubic centimeters per day |
| 1,000,000 | cubic centimeters per hour |
| 16,666.666680 | cubic centimeters per minute |
| 277.777778 | cubic centimeters per second |
| 24,000,000,000 | cubic millimeters per day |
| 1,000,000,000 | cubic millimeters per hour |
| 16,666,800 | cubic millimeters per minute |
| 277,777.777778 | cubic millimeters per second |

## CUBIC METERS PER MINUTE: =

| | |
|---|---|
| 1,440 | kiloliters per day |
| 60 | kiloliters per hour |
| 1 | kiloliters per minute |
| 0.016667 | kiloliters per second |
| 1,440 | cubic meters per day |
| 60 | cubic meters per hour |
| 1 | cubic meters per minute |
| 0.016667 | cubic meters per second |
| 1,883.43360 | cubic yards per day |
| 78.47640 | cubic yards per hour |
| 1.30794 | cubic yards per minute |
| 0.021799 | cubic yards per second |
| 9,057.51792 | barrels per day |
| 377.39658 | barrels per hour |
| 6.289943 | barrels per minute |
| 0.10483 | barrels per second |
| 14,400 | hektoliters per day |
| 600 | hektoliters per hour |
| 10 | hektoliters per minute |
| 0.16667 | hektoliters per second |
| 50,852.4480 | cubic feet per day |

CUBIC METERS PER MINUTE (cont'd): =

| | |
|---|---|
| 2,118.8520 | cubic feet per hour |
| 35.31444 | cubic feet per minute |
| 0.58857 | cubic feet per second |
| 144,000 | dekaliters per day |
| 6,000 | dekaliters per hour |
| 100 | dekaliters per minute |
| 1.66667 | dekaliters per second |
| 380,415.7728 | gallons per day |
| 15,850.65720 | gallons per hour |
| 264.17762 | gallons per minute |
| 4.40296 | gallons per second |
| 316,764.60480 | gallons (Imperial) per day |
| 13,198.5252 | gallons (Imperial) per hour |
| 219.97542 | gallons (Imperial) per minute |
| 3.66626 | gallons (Imperial) per second |
| 1,440,000 | liters per day |
| 60,000 | liters per hour |
| 1,000 | liters per minute |
| 16.66667 | liters per second |
| 1,440,000 | cubic decimeters per day |
| 60,000 | cubic decimeters per hour |
| 1,000 | cubic decimeters per minute |
| 16.66667 | cubic decimeters per second |
| 1,521,663.6672 | quarts per day |
| 63,402.65280 | quarts per hour |
| 1,056.71088 | quarts per minute |
| 17.61185 | quarts per second |
| 3,043,327.3344 | pints per day |
| 126,805.30560 | pints per hour |
| 2,113.42176 | pints per minute |
| 35.22370 | pints per second |
| 12,173,309.8560 | gills per day |
| 507,221.2440 | gills per hour |
| 8,453.68704 | gills per minute |
| 140.89479 | gills per second |
| 14,400,000 | deciliters per day |
| 600,000 | deciliters per hour |
| 10,000 | deciliters per minute |
| 166.66667 | deciliters per second |
| 87,873,031.8720 | cubic inches per day |
| 3,661,376.3280 | cubic inches per hour |
| 61,022.93879 | cubic inches per minute |
| 1,017.04898 | cubic inches per second |
| 144,000,000 | centiliters per day |

## CUBIC METERS PER MINUTE (cont'd): =

| | |
|---|---|
| 6,000,000 | centiliters per hour |
| 100,000 | centiliters per minute |
| 1,666.66667 | centiliters per second |
| 1,440,000,000 | milliliters per day |
| 60,000,000 | milliliters per hour |
| 1,000,000 | milliliters per minute |
| 16,666.66667 | milliliters per second |
| 1,440,000,000 | cubic centimeters per day |
| 60,000,000 | cubic centimeters per hour |
| 1,000,000 | cubic centimeters per minute |
| 16,666.66667 | cubic centimeters per second |
| 1,440,000,000,000 | cubic millimeters per day |
| 60,000,000,000 | cubic millimeters per hour |
| 1,000,000,000 | cubic millimeters per minute |
| 16,666,666.66667 | cubic millimeters per second |

## CUBIC METERS PER SECOND: =

| | |
|---|---|
| 86,400 | kiloliters per day |
| 3,600 | kiloliters per hour |
| 60 | kiloliters per minute |
| 1 | kiloliters per second |
| 86,400 | cubic meters per day |
| 3,600 | cubic meters per hour |
| 60 | cubic meters per minute |
| 1 | cubic meters per second |
| 113,006.0160 | cubic yards per day |
| 4,708.5840 | cubic yards per hour |
| 78.47640 | cubic yards per minute |
| 1.30794 | cubic yards per second |
| 543,451.07520 | barrels per day |
| 22,643.7948 | barrels per hour |
| 377.39658 | barrels per minute |
| 6.28994 | barrels per second |
| 864,000 | hektoliters per day |
| 36,000 | hektoliters per hour |
| 600 | hektoliters per minute |
| 10 | hektoliters per second |
| 3,051,167.616 | cubic feet per day |
| 127,131.9840 | cubic feet per hour |
| 2,118.86640 | cubic feet per minute |
| 35.31444 | cubic feet per second |
| 8,640,000 | dekaliters per day |

## CUBIC METERS PER SECOND (cont'd): =

| | |
|---|---|
| 360,000 | dekaliters per hour |
| 6,000 | dekaliters per minute |
| 100 | dekaliters per second |
| 22,824,946.3680 | gallons per day |
| 951,039.4320 | gallons per hour |
| 15,850.65720 | gallons per minute |
| 264.17762 | gallons per second |
| 19,005,798.5280 | gallons (Imperial) per day |
| 791,908.2720 | gallons (Imperial) per hour |
| 13,198.47120 | gallons (Imperial) per minute |
| 219.97542 | gallons (Imperial) per second |
| 86,400,000 | liters per day |
| 3,600,000 | liters per hour |
| 60,000 | liters per minute |
| 1,000 | liters per second |
| 86,400,000 | cubic decimeters per day |
| 3,600,000 | cubic decimeters per hour |
| 60,000 | cubic decimeters per minute |
| 1,000 | cubic decimeters per second |
| 91,299,820,0320 | quarts per day |
| 3,804,159.16800 | quarts per hour |
| 63,402.65280 | quarts per minute |
| 1,056.71088 | quarts per second |
| 182,599,640.0640 | pints per day |
| 7,608,318.3360 | pints per hour |
| 126,805.30560 | pints per minute |
| 2,113.42176 | pints per second |
| 730,398,560.2560 | gills per day |
| 30,433,273.3340 | gills per hour |
| 507,221.22240 | gills per minute |
| 8,453.68704 | gills per second |
| 864,000,000 | deciliters per day |
| 36,000,000 | deciliters per hour |
| 600,000 | deciliters per minute |
| 10,000 | deciliters per second |
| 5,272,381,911.4560 | cubic inches per day |
| 219,682,579.6440 | cubic inches per hour |
| 3,661,376.32740 | cubic inches per minute |
| 61,022.93879 | cubic inches per second |
| 8,640,000,000 | centiliters per day |
| 360,000,000 | centiliters per hour |
| 6,000,000 | centiliters per minute |
| 100,000 | centiliters per second |
| 86,400,000,000 | milliliters per day |

## CUBIC METERS PER SECOND (cont'd): =

| | |
|---|---|
| 3,600,000,000 | milliliters per hour |
| 60,000,000 | milliliters per minute |
| 1,000,000 | milliliters per second |
| 86,400,000,000 | cubic centimeters per day |
| 3,600,000,000 | cubic centimeters per hour |
| 60,000,000 | cubic centimeters per minute |
| 1,000,000 | cubic centimeters per second |
| 86,400,000,000,000 | cubic millimeters per day |
| 3,600,000,000,000 | cubic millimeters per hour |
| 60,000,000,000 | cubic millimeters per minute |
| 1,000,000,000 | cubic millimeters per second |

## CUBIC POISE CENTIMETER PER GRAM: =

| | |
|---|---|
| 0.0000000001 | square kilometers per second |
| 0.00000001 | square hektometers per second |
| 0.000001 | square dekameters per second |
| 0.0001 | square meters per second |
| 0.0010764 | square feet per second |
| 0.1550 | square inches per second |
| 0.01 | square decimeters per second |
| 1.0 | square centimeters per second |
| 100 | square millimeters per second |

## CUBIC POISE FOOT PER POUND: =

| | |
|---|---|
| 0.000000006243 | square kilometers per second |
| 0.0000006243 | square hektometers per second |
| 0.00006243 | square dekameters per second |
| 0.006243 | square meters per second |
| 0.067200 | square feet per second |
| 9.67680 | square inches per second |
| 0.6243 | square decimeters per second |
| 62.43 | square centimeters per second |
| 6,243 | square millimeters per second |

## CUBIC POISE INCH PER GRAM: =

| | |
|---|---|
| 0.0000000016387 | square kilometers per second |
| 0.00000016387 | square hektometers per second |
| 0.000016387 | square dekameters per second |
| 0.0016387 | square meters per second |

## CUBIC POISE INCH PER GRAM (cont'd); =

| | |
|---|---|
| 0.017639 | square feet per second |
| 2.540 | square inches per second |
| 0.16387 | square decimeters per second |
| 16.387 | square centimeters per second |
| 1,638.7 | square millimeters per second |

## CUBIC YARD: =

| | |
|---|---|
| 0.00000000047509 | cubic miles |
| 0.00000000076456 | cubic kilometers |
| 0.000000093914 | cubic furlongs |
| 0.00000076456 | cubic hektometers |
| 0.000093914 | cubic chains |
| 0.00076456 | cubic dekameters |
| 0.0060105 | cubic rods |
| 0.76456 | kiloliters |
| 0.76456 | cubic meters |
| 1.25971 | cubic Varas (Texas) |
| 4.80897 | barrels |
| 7.64559 | hektoliters |
| 21.69666 | bushels (U.S.) dry |
| 21.0220 | bushels (Imperial) dry |
| 27 | cubic feet |
| 63.99891 | cubic spans |
| 76.45595 | dekaliters |
| 86.78610 | pecks |
| 93.91329 | cubic links |
| 201.97350 | gallons (U.S.) liquid |
| 173.59299 | gallons (U.S.) dry |
| 168.18003 | gallons (Imperial) |
| 731.18592 | cubic hands |
| 764.55945 | liters |
| 764.55945 | cubic decimeters |
| 807.89400 | quarts (liquid) |
| 694.28070 | quarts (dry) |
| 1,615.78800 | pints (liquid) |
| 1,388.59218 | pints (dry) |
| 6,463.15200 | gills (liquid) |
| 7,645.5945 | deciliters |
| 46,656 | cubic inches |
| 76,455.945 | centiliters |
| 764,559.45 | milliliters |
| 764,559.45 | cubic centimeters |

## CUBIC YARD (cont'd) =

| | |
|---|---|
| 764,559,450 | cubic millimeters |
| 25,852.79808 | ounces (fluid) |
| 206,820.85509 | drams (fluid) |
| 24.5457 | sacks of cement (set) |
| 1,683.45 | pounds of water @ 60° F. |
| 1,736.10 | pounds of salt water |
| 1,944 | pounds of salt water @ 60° F. and 1.155 specific gravity |
| 13,217.634 | pounds of steel of 7.851 specific gravity |

## CUBIC YARD PER DAY: =

| | |
|---|---|
| 0.76456 | kiloliters per day |
| 0.031857 | kiloliters per hour |
| 0.00053095 | kiloliters per minute |
| 0.0000088491 | kiloliters per second |
| 0.76456 | cubic meters per day |
| 0.031857 | cubic meters per hour |
| 0.00053095 | cubic meters per minute |
| 0.0000088491 | cubic meters per second |
| 1.0 | cubic yards per day |
| 0.04166 | cubic yards per hour |
| 0.00069444 | cubic yards per minute |
| 0.000011574 | cubic yards per second |
| 4.80897 | barrels per day |
| 0.20037 | barrels per hour |
| 0.0033395 | barrels per minute |
| 0.000055659 | barrels per second |
| 7.64559 | hektoliters per day |
| 0.31857 | hektoliters per hour |
| 0.0053095 | hektoliters per minute |
| 0.00008849 | hektoliters per second |
| 27 | cubic feet per day |
| 1.1250 | cubic feet per hour |
| 0.018750 | cubic feet per minute |
| 0.00031250 | cubic feet per second |
| 76.45595 | dekaliters per day |
| 3.18568 | dekaliters per hour |
| 0.053095 | dekaliters per minute |
| 0.00088491 | dekaliters per second |
| 201.97350 | gallons per day |
| 8.41572 | gallons per hour |
| 0.14026 | gallons per minute |

CUBIC YARD PER DAY: = (cont'd)

| | |
|---|---|
| 0.0023377 | gallons per second |
| 168.18003 | gallons (Imperial) per day |
| 7.00740 | gallons (Imperial) per hour |
| 0.11679 | gallons (Imperial) per minute |
| 0.0019465 | gallons (Imperial) per second |
| 764.55945 | liters per day |
| 31.85676 | liters per hour |
| 0.53095 | liters per minute |
| 0.0088491 | liters per second |
| 764.55945 | cubic decimeters per day |
| 31.85676 | cubic decimeters per hour |
| 0.53095 | cubic decimeters per minute |
| 0.0088491 | cubic decimeters per second |
| 807.8940 | quarts per day |
| 33.66216 | quarts per hour |
| 0.56104 | quarts per minute |
| 0.0093506 | quarts per second |
| 1,615.7880 | pints per day |
| 67.32360 | pints per hour |
| 1.12206 | pints per minute |
| 0.018701 | pints per second |
| 6,463.152 | gills per day |
| 269.2980 | gills per hour |
| 4.48830 | gills per minute |
| 0.074805 | gills per second |
| 7,645.5945 | deciliters per day |
| 318.56760 | deciliters per hour |
| 5.30946 | deciliters per minute |
| 0.088491 | deciliters per second |
| 46,656 | cubic inches per day |
| 1,944.0 | cubic inches per hour |
| 32.40 | cubic inches per minute |
| 0.5400 | cubic inches per second |
| 76,455.945 | centiliters per day |
| 3,185.6760 | centiliters per hour |
| 53.09460 | centiliters per minute |
| 0.88491 | centiliters per second |
| 764,559.45 | milliliters per day |
| 31,856.760 | milliliters per hour |
| 530.946 | milliliters per minute |
| 8.84907 | milliliters per second |
| 764,559.45 | cubic centimeters per day |
| 31,856.760 | cubic centimeters per hour |
| 530.946 | cubic centimeters per minute |

## CUBIC YARD PER DAY (cont'd): =

8.84907 . . . . . . . . . . . . . . . . . . . . . . . . . . .cubic centimeters per second
764,559,450 . . . . . . . . . . . . . . . . . . . . . . . . cubic millimeters per day
31,856,760 . . . . . . . . . . . . . . . . . . . . . . . . cubic millimeters per hour
530,946 . . . . . . . . . . . . . . . . . . . . . . . . . .cubic millimeters per minute
8,849.07 . . . . . . . . . . . . . . . . . . . . . . . . . cubic millimeters per second

## CUBIC YARD PER HOUR: =

18.34963 . . . . . . . . . . . . . . . . . . . . . . . . . . . . . . .kiloliters per day
0.76456 . . . . . . . . . . . . . . . . . . . . . . . . . . . . . . . kiloliters per hour
0.012743 . . . . . . . . . . . . . . . . . . . . . . . . . . . . . .kiloliters per minute
0.00021238 . . . . . . . . . . . . . . . . . . . . . . . . . . . . kiloliters per second
18.34963 . . . . . . . . . . . . . . . . . . . . . . . . . . . . cubic meters per day
0.76456 . . . . . . . . . . . . . . . . . . . . . . . . . . . . .cubic meters per hour
0.012743 . . . . . . . . . . . . . . . . . . . . . . . . . . cubic meters per minute
0.00021238 . . . . . . . . . . . . . . . . . . . . . . . . . .cubic meters per second
24.0 . . . . . . . . . . . . . . . . . . . . . . . . . . . . . . . cubic yards per day
1.0 . . . . . . . . . . . . . . . . . . . . . . . . . . . . . . . .cubic yards per hour
0.016667 . . . . . . . . . . . . . . . . . . . . . . . . . . . cubic yards per minute
0.00027778 . . . . . . . . . . . . . . . . . . . . . . . . . . cubic yards per second
115.41312 . . . . . . . . . . . . . . . . . . . . . . . . . . . . . . . barrels per day
4.80897 . . . . . . . . . . . . . . . . . . . . . . . . . . . . . . . . barrels per hour
0.080148 . . . . . . . . . . . . . . . . . . . . . . . . . . . . . . barrels per minute
0.0013358 . . . . . . . . . . . . . . . . . . . . . . . . . . . . . barrels per second
183.49632 . . . . . . . . . . . . . . . . . . . . . . . . . . . . .hektoliters per day
7.64559 . . . . . . . . . . . . . . . . . . . . . . . . . . . . . . hektoliters per hour
0.12743 . . . . . . . . . . . . . . . . . . . . . . . . . . . . .hektoliters per minute
0.0021238 . . . . . . . . . . . . . . . . . . . . . . . . . . .hektoliters per second
648.0 . . . . . . . . . . . . . . . . . . . . . . . . . . . . . . . cubic feet per day
27.0 . . . . . . . . . . . . . . . . . . . . . . . . . . . . . . . . cubic feet per hour
0.450 . . . . . . . . . . . . . . . . . . . . . . . . . . . . . . cubic feet per minute
0.00750 . . . . . . . . . . . . . . . . . . . . . . . . . . . . cubic feet per second
1,834.9632 . . . . . . . . . . . . . . . . . . . . . . . . . . . . dekaliters per day
76.45595 . . . . . . . . . . . . . . . . . . . . . . . . . . . . .dekaliters per hour
1.27428 . . . . . . . . . . . . . . . . . . . . . . . . . . . . dekaliters per minute
0.021238 . . . . . . . . . . . . . . . . . . . . . . . . . . . dekaliters per second
4,847.3856 . . . . . . . . . . . . . . . . . . . . . . . . . . . . . . gallons per day
201.97350 . . . . . . . . . . . . . . . . . . . . . . . . . . . . . . gallons per hour
3.36624 . . . . . . . . . . . . . . . . . . . . . . . . . . . . . .gallons per minute
0.056104 . . . . . . . . . . . . . . . . . . . . . . . . . . . . . gallons per second
4,036.34880 . . . . . . . . . . . . . . . . . . . . . .gallons (Imperial) per day
168.18003 . . . . . . . . . . . . . . . . . . . . . . . gallons (Imperial) per hour
2.80302 . . . . . . . . . . . . . . . . . . . . . . . . .gallons (Imperial) per minute

**CUBIC YARD PER HOUR (cont'd): =**

| | |
|---|---|
| 0.046717 | gallons (Imperial) per second |
| 18,349.632 | liters per day |
| 764.55945 | liters per hour |
| 12.7428 | liters per minute |
| 0.21238 | liters per second |
| 18,349.632 | cubic decimeters per day |
| 764.55945 | cubic decimeters per hour |
| 12.7428 | cubic decimeters per minute |
| 0.21238 | cubic decimeters per second |
| 19,389.8880 | quarts per day |
| 807.8940 | quarts per hour |
| 13.4652 | quarts per minute |
| 0.22442 | quarts per second |
| 38,778.9120 | pints per day |
| 1,615.7780 | pints per hour |
| 26.92980 | pints per minute |
| 0.44883 | pints per second |
| 155,115.648 | gills per day |
| 6,463.152 | gills per hour |
| 107.7192 | gills per minute |
| 1.79532 | gills per second |
| 183,496.32 | deciliters per day |
| 7,645.5945 | deciliters per hour |
| 127.428 | deciliters per minute |
| 2.12378 | deciliters per second |
| 1,119,744 | cubic inches per day |
| 46,656 | cubic inches per hour |
| 777.6 | cubic inches per minute |
| 12.960 | cubic inches per second |
| 1,834,963.2 | centiliters per day |
| 76,455.945 | centiliters per hour |
| 1,274.28 | centiliters per minute |
| 21.23776 | centiliters per second |
| 18,349,632 | milliliters per day |
| 764,559.45 | milliliters per hour |
| 12,742.8 | milliliters per minute |
| 212.37763 | milliliters per second |
| 18,349,632 | cubic centimeters per day |
| 764,559.45 | cubic centimeters per hour |
| 12,742.8 | cubic centimeters per minute |
| 212.37763 | cubic centimeters per second |
| 18,349,632,000 | cubic millimeters per day |
| 764,559,450 | cubic millimeters per hour |
| 12,742,800 | cubic millimeters per minute |
| 212,377.63 | cubic millimeters per second |

## CUBIC YARD PER MINUTE: =

| | |
|---|---|
| 1,100.96640 | kiloliters per day |
| 45.87360 | kiloliters per hour |
| 0.76456 | kiloliters per minute |
| 0.012743 | kiloliters per second |
| 1,100.96640 | cubic meters per day |
| 45.87360 | cubic meters per hour |
| 0.76456 | cubic meters per minute |
| 0.012743 | cubic meters per second |
| 1,440 | cubic yards per day |
| 60 | cubic yards per hour |
| 1.0 | cubic yards per minute |
| 0.016667 | cubic yards per second |
| 6,924.960 | barrels per day |
| 288.540 | barrels per hour |
| 4.80897 | barrels per minute |
| 0.080150 | barrels per second |
| 11,009.66399 | hektoliters per day |
| 458.73560 | hektoliters per hour |
| 7.64559 | hektoliters per minute |
| 0.12743 | hektoliters per second |
| 38,880 | cubic feet per day |
| 1,620 | cubic feet per hour |
| 27.0 | cubic feet per minute |
| 0.450 | cubic feet per second |
| 110,096.633994 | dekaliters per day |
| 4,587.3600 | dekaliters per hour |
| 76.45595 | dekaliters per minute |
| 1.27427 | dekaliters per second |
| 290,842.2720 | gallons per day |
| 12,118.4280 | gallons per hour |
| 201.97350 | gallons per minute |
| 3.36623 | gallons per second |
| 242,179.20 | gallons (Imperial) per day |
| 10,090.80 | gallons (Imperial) per hour |
| 168.18003 | gallons (Imperial) per minute |
| 2.80300 | gallons (Imperial) per second |
| 1,100,966.39942 | liters per day |
| 45,873.59998 | liters per hour |
| 764.55945 | liters per minute |
| 12.74267 | liters per second |
| 1,100,966.39942 | cubic decimeters per day |
| 45,873.59998 | cubic decimeters per hour |
| 764.55945 | cubic decimeters per minute |
| 12.74267 | cubic decimeters per second |

CUBIC YARD PER MINUTE (cont'd): =

| | |
|---|---|
| 1,163,367.36 | quarts per day |
| 48,473.64 | quarts per hour |
| 807.8940 | quarts per minute |
| 13.4649 | quarts per second |
| 2,326,734.72 | pints per day |
| 96,947.28 | pints per hour |
| 1,615.7880 | pints per minute |
| 26.9298 | pints per second |
| 9,306,938.88 | gills per day |
| 387,789.12 | gills per hour |
| 6,463.152 | gills per minute |
| 107.71920 | gills per second |
| 11,009,663.99424 | deciliters per day |
| 458,735.99976 | deciliters per hour |
| 7,645.5945 | deciliters per minute |
| 127.42667 | deciliters per second |
| 67,184,640 | cubic inches per day |
| 2,799,360.0 | cubic inches per hour |
| 46,656.0 | cubic inches per minute |
| 777.6 | cubic inches per second |
| 110,096.639.9424 | centiliters per day |
| 4,587,359.9976 | centiliters per hour |
| 76,455.945 | centiliters per minute |
| 1,274.26667 | centiliters per second |
| 1,100,966,399.424 | milliliters per day |
| 45,873,599.976 | milliliters per hour |
| 764,559.45 | milliliters per minute |
| 12,742.66667 | milliliters per second |
| 1,100,966,399.424 | cubic centimeters per day |
| 45,873,599.976 | cubic centimeters per hour |
| 764,559.45 | cubic centimeters per minute |
| 12,742.66667 | cubic centimeters per second |
| 1,100,966,399,424 | cubic millimeters per day |
| 45,873,599,976 | cubic millimeters per hour |
| 764,559,450 | cubic millimeters per minute |
| 12,742,666.67 | cubic millimeters per second |

CUBIC YARD PER SECOND: =

| | |
|---|---|
| 66,057.93648 | kiloliters per day |
| 2,752.41402 | kiloliters per hour |
| 45.87357 | kiloliters per minute |
| 0.76456 | kiloliters per second |
| 66,057.93648 | cubic meters per day |

CUBIC YARD PER SECOND (cont'd): =

| | |
|---|---|
| 2,752.41402 | cubic meters per hour |
| 45.87357 | cubic meters per minute |
| 0.76456 | cubic meters per second |
| 86,400 | cubic yards per day |
| 3,600 | cubic yards per hour |
| 60.0 | cubic yards per minute |
| 1.0 | cubic yards per second |
| 415,495.008 | barrels per day |
| 17,312.2920 | barrels per hour |
| 288.53820 | barrels per minute |
| 4.80897 | barrels per second |
| 660,579.3648 | hektoliters per day |
| 27,524.1402 | hektoliters per hour |
| 458.73567 | hektoliters per minute |
| 7.64559 | hektoliters per second |
| 2,332,800 | cubic feet per day |
| 97,200 | cubic feet per hour |
| 1,620 | cubic feet per minute |
| 27 | cubic feet per second |
| 6,605,793.648 | dekaliters per day |
| 275,241.402 | dekaliters per hour |
| 4,587.3567 | dekaliters per minute |
| 76.45595 | dekaliters per second |
| 17,450,510.40 | gallons per day |
| 727,104.60 | gallons per hour |
| 12,118.410 | gallons per minute |
| 201.97350 | gallons per second |
| 14,530,754.5920 | gallons (Imperial) per day |
| 605,448.1080 | gallons (Imperial) per hour |
| 10,090.80180 | gallons (Imperial) per minute |
| 168.18003 | gallons (Imperial) per second |
| 66,057,936.48 | liters per day |
| 2,752,414.02 | liters per hour |
| 45,873.567 | liters per minute |
| 764.55945 | liters per second |
| 66,057,936.48 | cubic decimeters per day |
| 2,752,414.02 | cubic decimeters per hour |
| 45,873.567 | cubic decimeters per minute |
| 764.55945 | cubic decimeters per second |
| 69,802,041.6 | quarts per day |
| 2,908.418.40 | quarts per hour |
| 48,473.640 | quarts per minute |
| 807.8940 | quarts per second |
| 139,604,083.2 | pints per day |

**CUBIC YARD PER SECOND (cont'd): =**

5,816,836.8 . . . . . . . . . . . . . . . . . . . . . . pints per hour
96,947.280 . . . . . . . . . . . . . . . . . . . . . . pints per minute
1,615.7880 . . . . . . . . . . . . . . . . . . . . . . pints per second
558,416,332.8 . . . . . . . . . . . . . . . . . . . . .gills per day
23,267,347.20 . . . . . . . . . . . . . . . . . . . . gills per hour
387,789.120 . . . . . . . . . . . . . . . . . . . . .gills per minute
6,463.152 . . . . . . . . . . . . . . . . . . . . . .gills per second
660,579,364.8 . . . . . . . . . . . . . . . . . . .deciliters per day
27,524,140.2 . . . . . . . . . . . . . . . . . . . deciliters per hour
458,735.67 . . . . . . . . . . . . . . . . . . . .deciliters per minute
7,645.5945 . . . . . . . . . . . . . . . . . . . .deciliters per second
4,031,078,400 . . . . . . . . . . . . . . . . . . .cubic inches per day
167,961,600 . . . . . . . . . . . . . . . . . . . cubic inches per hour
2,799,360 . . . . . . . . . . . . . . . . . . . .cubic inches per minute
46,656 . . . . . . . . . . . . . . . . . . . . . cubic inches per second
6,605,793,648 . . . . . . . . . . . . . . . . . . . centiliters per day
275,241,402 . . . . . . . . . . . . . . . . . . . centiliters per hour
4,587,356.7 . . . . . . . . . . . . . . . . . . . centiliters per minute
76,445.945 . . . . . . . . . . . . . . . . . . . . centiliters per second
66,057,936,480 . . . . . . . . . . . . . . . . . . milliliters per day
2,752,414,020 . . . . . . . . . . . . . . . . . . .milliliters per hour
45,873,567 . . . . . . . . . . . . . . . . . . . . milliliters per minute
764,559.45 . . . . . . . . . . . . . . . . . . . . milliliters per second
66,057,936,480 . . . . . . . . . . . . . . . . .cubic centimeters per day
2,752,414,020 . . . . . . . . . . . . . . . . . cubic centimeters per hour
45,873,565 . . . . . . . . . . . . . . . . . .cubic centimeters per minute
764,559,45 . . . . . . . . . . . . . . . . . .cubic centimeters per second
66,057,936,480,000 . . . . . . . . . . . . . . . . cubic millimeters per day
2,752,414,020,000 . . . . . . . . . . . . . . . cubic millimeters per hour
45,873,565,000 . . . . . . . . . . . . . . . .cubic millimeters per minute
764,559,450 . . . . . . . . . . . . . . . . . . cubic millimeters per second

**DECIGRAM: =**

0.0000001102311150 . . . . . . . . . . . . . . . . . . . .tons (short)
0.0000000984206383 . . . . . . . . . . . . . . . . . . . . tons (long)
0.0000001 . . . . . . . . . . . . . . . . . . . . . . . . tons (metric)
0.0001 . . . . . . . . . . . . . . . . . . . . . . . . . . . kilograms
0.000267922895 . . . . . . . . . . . . . . . . . . . . . pounds (Troy)
0.00022046223 . . . . . . . . . . . . . . . . . . . . . pounds (Avoir.)
0.001 . . . . . . . . . . . . . . . . . . . . . . . . . . . . hektograms
0.01 . . . . . . . . . . . . . . . . . . . . . . . . . . . . . dekagrams
0.00321507 . . . . . . . . . . . . . . . . . . . . . . . . ounces (Troy)

**DECIGRAM (cont'd): =**

0.00352739568 . . . . . . . . . . . . . . . . . . . . . . . . .ounces (Avoir.)
0.1 . . . . . . . . . . . . . . . . . . . . . . . . . . . . . . . . . . . . . . . grams
10 . . . . . . . . . . . . . . . . . . . . . . . . . . . . . . . . . . . . centigrams
100 . . . . . . . . . . . . . . . . . . . . . . . . . . . . . . . . . . . .milligrams
1.543236 . . . . . . . . . . . . . . . . . . . . . . . . . . . . . . . . . . . grains
0.0257205 . . . . . . . . . . . . . . . . . . . . . . . . . . . . . .drams (Troy)
0.05643833088 . . . . . . . . . . . . . . . . . . . . . . . . . . drams (Avoir.)
0.5 . . . . . . . . . . . . . . . . . . . . . . . . . . . . . . . . . . carats (metric)

**DECILITER: =**

0.0001 . . . . . . . . . . . . . . . . . . . . . . . . . . . . . . . . . . kiloliters
0.0001 . . . . . . . . . . . . . . . . . . . . . . . . . . . . . . . cubic meters
0.0001308 . . . . . . . . . . . . . . . . . . . . . . . . . . . . . . cubic yards
0.00062900 . . . . . . . . . . . . . . . . . . . . . . . . . . . . . . . . barrels
0.001 . . . . . . . . . . . . . . . . . . . . . . . . . . . . . . . . . hektoliters
0.0028380 . . . . . . . . . . . . . . . . . . . . . . . . . . . . bushels (U.S.)
0.0027513 . . . . . . . . . . . . . . . . . . . . . . . . .bushels (Imperial) dry
0.0035316 . . . . . . . . . . . . . . . . . . . . . . . . . . . . . . . . cubic feet
0.01 . . . . . . . . . . . . . . . . . . . . . . . . . . . . . . . . . . . . dekaliters
0.011352 . . . . . . . . . . . . . . . . . . . . . . . . . . . . . . . . . . . pecks
0.026418 . . . . . . . . . . . . . . . . . . . . . . . . . . gallons (U.S.) liquid
0.022706 . . . . . . . . . . . . . . . . . . . . . . . . . . . . gallons (U.S.) dry
0.021997 . . . . . . . . . . . . . . . . . . . . . . . . . . . gallons (Imperial)
0.10567 . . . . . . . . . . . . . . . . . . . . . . . . . . . . . . . quarts (liquid)
0.090816 . . . . . . . . . . . . . . . . . . . . . . . . . . . . . . . .quarts (dry)
0.1 . . . . . . . . . . . . . . . . . . . . . . . . . . . . . . . . . . . . . . . . .liters
0.1 . . . . . . . . . . . . . . . . . . . . . . . . . . . . . . . . cubic decimeters
0.18163 . . . . . . . . . . . . . . . . . . . . . . . . . . . . . . . . . pints (dry)
0.21134 . . . . . . . . . . . . . . . . . . . . . . . . . . . . . . . . pints (liquid)
0.84536 . . . . . . . . . . . . . . . . . . . . . . . . . . . . . . . . .gills (liquid)
6.1025 . . . . . . . . . . . . . . . . . . . . . . . . . . . . . . . . . cubic inches
1.0 . . . . . . . . . . . . . . . . . . . . . . . . . . . . . . . . . . . . . centiliters
100 . . . . . . . . . . . . . . . . . . . . . . . . . . . . . . . . . . . . . milliliters
100 . . . . . . . . . . . . . . . . . . . . . . . . . . . . . . . cubic centimeters
100,000 . . . . . . . . . . . . . . . . . . . . . . . . . . . . .cubic millimeters
3.38147 . . . . . . . . . . . . . . . . . . . . . . . . . . . . . . . . ounces (fluid)
27.05179 . . . . . . . . . . . . . . . . . . . . . . . . . . . . . . . . drams (fluid)

**DECIMETER: =**

0.000062137 . . . . . . . . . . . . . . . . . . . . . . . . . . . . . . . . . . .miles
0.0001 . . . . . . . . . . . . . . . . . . . . . . . . . . . . . . . . . . kilometers

DECIMETER (cont'd): =

| | |
|---|---|
| 0.00049710 | furlongs |
| 0.001 | hektometers |
| 0.0049710 | chains |
| 0.01 | dekameters |
| 0.019884 | rods |
| 0.1 | meters |
| 0.10935 | yards |
| 0.11811 | varas (Texas) |
| 0.3280833 | feet |
| 0.43744 | spans |
| 0.49710 | links |
| 0.98425 | hands |
| 1 | decimeter |
| 10 | centimeter |
| 3.93700 | inches |
| 100 | millimeters |
| 3,937 | mils |
| 100,000 | microns |
| 100,000,000 | millimicrons |
| 100,000,000 | micro millimeters |
| 155,316 | wave lengths of red line of cadmium |
| 100,000,000,000 | Angstrom units |

DEGREE (ANGLE): =

| | |
|---|---|
| 0.0027778 | circumferences |
| 0.0027778 | revolutions |
| 0.01111 | quadrants |
| 0.017453 | radians |
| 1 | hours |
| 60 | minutes |
| 3,600 | seconds |

DEGREE PER DAY: =

| | |
|---|---|
| 0.0027778 | revolutions per day |
| 0.00011574 | revolutions per hour |
| 0.0000019290 | revolutions per minute |
| 0.000000032150 | revolutions per second |
| 0.01111 | quadrants per day |
| 0.00046292 | quadrants per hour |

**DEGREE PER DAY (cont'd): =**

| | |
|---|---|
| 0.0000077154 | quadrants per minute |
| 0.00000012859 | quadrants per second |
| 0.017453 | radians per day |
| 0.00072720 | radians per hour |
| 0.000012120 | radians per minute |
| 0.00000020200 | radians per second |
| 1.0 | degrees per day |
| 0.041666 | degrees per hour |
| 0.00069444 | degrees per minute |
| 0.000011574 | degrees per second |
| 1.0 | hours per day |
| 0.041666 | hours per hour |
| 0.00069444 | hours per minute |
| 0.000011574 | hours per second |
| 60 | minutes per day |
| 2.5 | minutes per hour |
| 0.041666 | minutes per minute |
| 0.00069444 | minutes per second |
| 3,600 | seconds per day |
| 150 | seconds per hour |
| 2.5 | seconds per minute |
| 0.041667 | seconds per second |

**DEGREE PER HOUR: =**

| | |
|---|---|
| 0.066667 | revolutions per day |
| 0.0027778 | revolutions per hour |
| 0.000046297 | revolutions per minute |
| 0.00000077161 | revolutions per second |
| 0.26664 | quadrants per day |
| 0.01111 | quadrants per hour |
| 0.00018517 | quadrants per minute |
| 0.0000030861 | quadrants per second |
| 0.41888 | radians per day |
| 0.017453 | radians per hour |
| 0.00029088 | radians per minute |
| 0.0000048481 | radians per second |
| 24 | degrees per day |
| 1.0 | degrees per hour |
| 0.016667 | degrees per minute |
| 0.00027778 | degrees per second |
| 24 | hours per day |
| 1.0 | hours per hour |

DEGREE PER HOUR (cont'd): =

0.016667 . . . . . . . . . . . . . . . . . . . . . . . . . . . . . .hours per minute
0.00027778 . . . . . . . . . . . . . . . . . . . . . . . . . . .hours per second
1,440 . . . . . . . . . . . . . . . . . . . . . . . . . . . . . . . minutes per day
60 . . . . . . . . . . . . . . . . . . . . . . . . . . . . . . . . .minutes per hour
1 . . . . . . . . . . . . . . . . . . . . . . . . . . . . . . . . minutes per minute
0.016667 . . . . . . . . . . . . . . . . . . . . . . . . . . . minutes per second
86,400 . . . . . . . . . . . . . . . . . . . . . . . . . . . . . . . seconds per day
3,600 . . . . . . . . . . . . . . . . . . . . . . . . . . . . . . .seconds per hour
60 . . . . . . . . . . . . . . . . . . . . . . . . . . . . . . . . seconds per minute
1 . . . . . . . . . . . . . . . . . . . . . . . . . . . . . . . . . second per second

DEGREE PER MINUTE: =

4.0 . . . . . . . . . . . . . . . . . . . . . . . . . . . . . . . revolutions per day
0.16667 . . . . . . . . . . . . . . . . . . . . . . . . . . . .revolutions per hour
0.0027778 . . . . . . . . . . . . . . . . . . . . . . . . revolutions per minute
0.000046297 . . . . . . . . . . . . . . . . . . . . . . . revolutions per second
15.99869 . . . . . . . . . . . . . . . . . . . . . . . . . . . . quadrants per day
0.66661 . . . . . . . . . . . . . . . . . . . . . . . . . . . . quadrants per hour
0.01111 . . . . . . . . . . . . . . . . . . . . . . . . . . . quadrants per minute
0.00018517 . . . . . . . . . . . . . . . . . . . . . . . . quadrants per second
25.13203 . . . . . . . . . . . . . . . . . . . . . . . . . . . . . .radians per day
1.04717 . . . . . . . . . . . . . . . . . . . . . . . . . . . . . . radians per hour
0.017453 . . . . . . . . . . . . . . . . . . . . . . . . . . . .radians per minute
0.00029088 . . . . . . . . . . . . . . . . . . . . . . . . . . radians per second
1,440 . . . . . . . . . . . . . . . . . . . . . . . . . . . . . . .degrees per day
60 . . . . . . . . . . . . . . . . . . . . . . . . . . . . . . . . . degrees per hour
1.0 . . . . . . . . . . . . . . . . . . . . . . . . . . . . . . .degrees per minute
0.016667 . . . . . . . . . . . . . . . . . . . . . . . . . . .degrees per second
1,440 . . . . . . . . . . . . . . . . . . . . . . . . . . . . . . . . .hours per day
60 . . . . . . . . . . . . . . . . . . . . . . . . . . . . . . . . . . hours per hour
1.0 . . . . . . . . . . . . . . . . . . . . . . . . . . . . . . . . .hours per minute
0.016667 . . . . . . . . . . . . . . . . . . . . . . . . . . . . .hours per second
86,400 . . . . . . . . . . . . . . . . . . . . . . . . . . . . . . . minutes per day
3,600 . . . . . . . . . . . . . . . . . . . . . . . . . . . . . . .minutes per hour
60 . . . . . . . . . . . . . . . . . . . . . . . . . . . . . . . . minutes per minute
1 . . . . . . . . . . . . . . . . . . . . . . . . . . . . . . . . minutes per second
5,184,000 . . . . . . . . . . . . . . . . . . . . . . . . . . . . seconds per day
216,000 . . . . . . . . . . . . . . . . . . . . . . . . . . . . .seconds per hour
3,600 . . . . . . . . . . . . . . . . . . . . . . . . . . . . . . seconds per minute
60 . . . . . . . . . . . . . . . . . . . . . . . . . . . . . . . . seconds per second

## EGREE PER SECOND: =

| | |
|---|---|
| 240 | revolutions per day |
| 10 | revolutions per hour |
| 0.16667 | revolutions per minute |
| 0.0027778 | revolutions per second |
| 959.904 | quadrants per day |
| 39.9960 | quadrants per hour |
| 0.6666 | quadrants per minute |
| 0.01111 | quadrants per second |
| 1,507.9392 | radians per day |
| 62.83080 | radians per hour |
| 1.047180 | radians per minute |
| 0.017453 | radians per second |
| 86,400 | degrees per day |
| 3,600 | degrees per hour |
| 60 | degrees per minute |
| 1.0 | degrees per second |
| 86,400 | hours per day |
| 3,600 | hours per hour |
| 60 | hours per minute |
| 1.0 | hours per second |
| 5,184,000 | minutes per day |
| 216,000 | minutes per hour |
| 3,600 | minutes per minute |
| 60 | minutes per second |
| 311,040,000 | seconds per day |
| 12,960,000 | seconds per hour |
| 216,000 | seconds per minute |
| 3,600 | seconds per second |

## DEKAGRAM: =

| | |
|---|---|
| 0.00000984206383 | tons (long) |
| 0.00001 | tons (metric) |
| 0.0000110231115 | tons (short) |
| 0.01 | kilograms |
| 0.022046223 | pounds (Avoir.) |
| 0.0267922895 | pounds (Troy) |
| 0.1 | hektograms |
| 0.321507 | ounces (Troy) |
| 0.352739568 | ounces (Avoir.) |
| 2.572053 | drams (Troy) |
| 5.6438330880 | drams (Avoir.) |
| 6.430149 | pennyweights |

DEKAGRAM (cont'd): =

| | |
|---|---|
| 7.71618 | scruples |
| 10 | grams |
| 100 | decigrams |
| 154.3234765625 | grains |
| 1,000 | centigrams |
| 10,000 | milligrams |
| 50 | carats (metric) |

DEKALITER: =

| | |
|---|---|
| 0.01 | kiloliters |
| 0.01 | cubic meters |
| 0.01308 | cubic yards |
| 0.06290 | barrels |
| 0.1 | hektoliters |
| 0.28380 | bushels (U.S.) |
| 0.27513 | bushels (Imperial) dry |
| 0.35316 | cubic feet |
| 1.1352 | pecks |
| 2.6418 | gallons (U.S.) liquid |
| 2.2706 | gallons (U.S.) dry |
| 2.1997 | gallons (Imperial) |
| 10.567 | quarts (liquid) |
| 9.0816 | quarts (dry) |
| 10 | liters |
| 10 | cubic decimeters |
| 18.163 | pints (dry) |
| 21.134 | pints (liquid) |
| 84.536 | gills (liquid) |
| 100 | deciliters |
| 610.25 | cubic inches |
| 1,000 | centiliters |
| 10,000 | milliliters |
| 10,000 | cubic centimeters |
| 100,000,000 | cubic millimeters |
| 338.147 | ounces (fluid) |
| 2,705.179 | drams (fluid) |

DEKAMETER: =

| | |
|---|---|
| 0.0062137 | miles |
| 0.01 | kilometers |

DEKAMETER: = (cont'd)

| | |
|---|---|
| 0.049710 | furlongs |
| 0.1 | hektometers |
| 0.49710 | chains |
| 1.0 | dekameters |
| 1.98838 | rods |
| 10 | meters |
| 10.935 | yards |
| 11.811 | varas (Texas) |
| 32.80830 | feet |
| 43.744 | spans |
| 49.710 | links |
| 98.425 | hands |
| 100 | decimeters |
| 1000 | centimeters |
| 393.70 | inches |
| 10,000 | millimeters |
| 393,700 | mils |
| 10,000,000 | microns |
| 10,000,000,000 | millimicrons |
| 10,000,000,000 | micro millimeters |
| 15,531,595 | wave lengths of red line of cadmium |
| 10,000,000,000 x $10^3$ | Angstrom Units |

DRAM (AVOIRDUPOIS): =

| | |
|---|---|
| 0.0000017439 | tons (long) |
| 0.0000017718 | tons (metric) |
| 0.0000019531 | tons (short) |
| 0.001771845 | kilograms |
| 0.00390625 | pounds (Avoir.) |
| 0.0047471788 | pounds (Troy) |
| 0.01771845 | hektograms |
| 0.1771845 | dekagrams |
| 0.056966146 | ounces (Troy) |
| 0.0625 | ounces (Avoir.) |
| 0.4557292 | drams (Troy) |
| 1.139323 | pennyweights |
| 1.3671875 | scruples |
| 1.771845 | grams |
| 17.71845 | decigrams |
| 27.34375 | grains |
| 177.1845 | centigrams |

## DRAM (AVOIRDUPOIS) (cont'd): =

| | |
|---|---|
| 1,771.845 | milligrams |
| 8.85923 | carats (metric) |

## DRAM (FLUID): =

| | |
|---|---|
| 0.0000036966 | kiloliters |
| 0.0000036966 | cubic meters |
| 0.0000048352 | cubic yards |
| 0.000023252 | barrels |
| 0.000036966 | hektoliters |
| 0.00013055 | cubic feet |
| 0.00036966 | dekaliters |
| 0.00097658 | gallons (U.S.) liquid |
| 0.00081318 | gallons (Imperial) liquid |
| 0.00390625 | quarts (liquid) |
| 0.0036966 | liters |
| 0.0036966 | cubic decimeters |
| 0.0078125 | pints (liquid) |
| 0.03125 | gills (liquid) |
| 0.036966 | deciliters |
| 0.225586 | cubic inches |
| 0.36966 | centiliters |
| 3.69661 | milliliters |
| 3.69661 | cubic centimeters |
| 3,696.61 | cubic millimeters |
| 0.125 | ounces (fluid) |
| 60 | minims |

## DRAM (TROY OR APOTHECARY): =

| | |
|---|---|
| 0.0000038265308 | tons (long) |
| 0.0000038879351 | tons (metric) |
| 0.0000042857145 | tons (short) |
| 0.0038879351 | kilograms |
| 0.008571429 | pounds (Avoir.) |
| 0.010416667 | pounds (Troy) |
| 0.038879351 | hektograms |
| 0.38879351 | dekagrams |
| 0.12500 | ounces (Troy) |
| 0.1371429 | ounces (Avoir.) |
| 2.194286 | drams (Avoir.) |
| 2.50 | pennyweights |

**DRAM (TROY OR APOTHECARY) (cont'd): =**

3.0 . . . . . . . . . . scruples
3.8879351 . . . . . . . . . . grams
38.879351 . . . . . . . . . . decigrams
60.0 . . . . . . . . . . grains
388.79351 . . . . . . . . . . centigrams
3,887.9351 . . . . . . . . . . milligrams
19.43968 . . . . . . . . . . carats (metric)

**FOOT (OR ENGINEER'S LINK): =**

0.0001893939 . . . . . . . . . . miles
0.0003048006 . . . . . . . . . . kilometers
0.00151515 . . . . . . . . . . furlongs
0.003048006 . . . . . . . . . . hektometers
0.0151515 . . . . . . . . . . chains
0.03048006 . . . . . . . . . . dekameters
0.0606061 . . . . . . . . . . rods
0.3048006 . . . . . . . . . . meters
0.33333 . . . . . . . . . . yards
0.3600 . . . . . . . . . . varas (Texas)
1.33333 . . . . . . . . . . spans
1.515152 . . . . . . . . . . links
3.00 . . . . . . . . . . hands
3.048006 . . . . . . . . . . decimeters
30.48006 . . . . . . . . . . centimeters
12 . . . . . . . . . . inches
304.8006 . . . . . . . . . . millimeters
12,000 . . . . . . . . . . mils
304,801 . . . . . . . . . . microns
304,801,200 . . . . . . . . . . millimicrons
304,801,200 . . . . . . . . . . micro millimeters
473,404 . . . . . . . . . . wave lengths of red line of cadmium
3,048,012,000 . . . . . . . . . . Angstrom Units

**FOOT PER DAY: =**

0.0001893939 . . . . . . . . . . miles per day
0.0000078914 . . . . . . . . . . miles per hour
0.00000013152 . . . . . . . . . . miles per minute
0.0000000021921 . . . . . . . . . . miles per secoond
0.0003048006 . . . . . . . . . . kilometers per day

FOOT PER DAY (cont'd): =

| | |
|---|---|
| 0.0000127 | kilometers per hour |
| 0.00000021167 | kilometers per minute |
| 0.0000000035278 | kilometers per second |
| 0.00151515 | furlongs per day |
| 0.000063131 | furlongs per hour |
| 0.0000010522 | furlongs per minute |
| 0.000000017536 | furlongs per second |
| 0.003048006 | hektometers per day |
| 0.000127 | hektometers per hour |
| 0.0000021167 | hektometers per minute |
| 0.000000035278 | hektometers per second |
| 0.0151515 | chains per day |
| 0.00063131 | chains per hour |
| 0.000010522 | chains per minute |
| 0.00000017536 | chains per second |
| 0.03048006 | dekameters per day |
| 0.00127 | dekameters per hour |
| 0.000021167 | dekameters per minute |
| 0.00000035278 | dekameters per second |
| 0.0606061 | rods per day |
| 0.0025253 | rods per hour |
| 0.000042088 | rods per minute |
| 0.00000070146 | rods per second |
| 0.3048006 | meters per day |
| 0.0127 | meters per hour |
| 0.00021167 | meters per minute |
| 0.0000035278 | meters per second |
| 0.33333 | yards per day |
| 0.013889 | yards per hour |
| 0.00023148 | yards per minute |
| 0.0000038580 | yards per second |
| 0.3600 | varas (Texas) per day |
| 0.015 | varas (Texas) per hour |
| 0.00025 | varas (Texas) per minute |
| 0.0000041667 | varas (Texas) per second |
| 1.0 | feet per day |
| 0.041667 | feet per hour |
| 0.00069444 | feet per minute |
| 0.000011574 | feet per second |
| 1.33333 | spans per day |
| 0.055556 | spans per hour |
| 0.00092593 | spans per minute |
| 0.000015432 | spans per second |
| 1.515152 | links per day |

FOOT PER DAY (cont'd): =

| | |
|---|---|
| 0.063131 | links per hour |
| 0.0010522 | links per minute |
| 0.000017536 | links per second |
| 3.00 | hands per day |
| 0.125 | hands per hour |
| 0.0020833 | hands per minute |
| 0.000034722 | hands per second |
| 3.048006 | decimeters per day |
| 0.127 | decimeters per hour |
| 0.0021167 | decimeters per minute |
| 0.000035278 | decimeters per second |
| 30.48006 | centimeters per day |
| 1.270 | centimeters per hour |
| 0.021167 | centimeters per minute |
| 0.00035278 | centimers per second |
| 12.0 | inches per day |
| 0.50 | inches per hour |
| 0.0083333 | inches per minute |
| 0.00013889 | inches per second |
| 304.8006 | millimeters per day |
| 12.70004 | millimeters per hour |
| 0.21167 | millimeters per minute |
| 0.0035278 | millimeters per second |

FOOT PER HOUR: =

| | |
|---|---|
| 0.0045455 | miles per day |
| 0.0001893939 | miles per hour |
| 0.0000031566 | miles per minute |
| 0.000000052609 | miles per second |
| 0.0073152 | kilometers per day |
| 0.0003048006 | kilometers per hour |
| 0.00000508 | kilometers per minute |
| 0.000000084667 | kilometers per second |
| 0.036364 | furlongs per day |
| 0.00151515 | furlongs per hour |
| 0.000025253 | furlongs per minute |
| 0.00000042088 | furlongs per second |
| 0.073152 | hektometers per day |
| 0.003048006 | hektometers per hour |
| 0.0000508 | hektometers per minute |
| 0.00000084667 | hektometers per second |
| 0.36364 | chains per day |

FOOT PER HOUR (cont'd): =

| | |
|---|---|
| 0.0151515 | chains per hour |
| 0.00025253 | chains per minute |
| 0.0000042088 | chains per second |
| 0.73152 | dekameters per day |
| 0.03048006 | dekameters per hour |
| 0.000508 | dekameters per minute |
| 0.0000084667 | dekameters per second |
| 1.45456 | rods per day |
| 0.0606061 | rods per hour |
| 0.00101010 | rods per minute |
| 0.000016835 | rods per second |
| 7.31521 | meters per day |
| 0.3048006 | meters per hour |
| 0.00508 | meters per minute |
| 0.000084667 | meters per second |
| 8.0000 | yards per day |
| 0.33333 | yards per hour |
| 0.0055556 | yards per minute |
| 0.000092593 | yards per second |
| 8.640 | varas (Texas) per day |
| 0.3600 | varas (Texas) per hour |
| 0.006 | varas (Texas) per minute |
| 0.0001 | varas (Texas) per second |
| 24.0 | feet per day |
| 1.0 | feet per hour |
| 0.016667 | feet per minute |
| 0.00027778 | feet per second |
| 32.000 | spans per day |
| 1.33333 | spans per hour |
| 0.022222 | spans per minute |
| 0.00037037 | spans per second |
| 36.36364 | links per day |
| 1.51515 | links per hour |
| 0.025253 | links per minute |
| 0.00042088 | links per second |
| 72.0 | hands per day |
| 3.0 | hands per hour |
| 0.05000 | hands per minute |
| 0.00083333 | hands per second |
| 73.15214 | decimeters per day |
| 3.048006 | decimeters per hour |
| 0.0508 | decimeters per minute |
| 0.00084667 | decimeters per second |
| 731.52144 | centimeters per day |

## FOOT PER HOUR (cont'd): =

| | |
|---|---|
| 30.48006 | centimeters per hour |
| 0.5080 | centimeters per minute |
| 0.0084667 | centimeters per second |
| 288.0 | inches per day |
| 12.0 | inches per hour |
| 0.200 | inches per minute |
| 0.0033333 | inches per second |
| 7,315.21440 | millimeters per day |
| 304.8006 | millimeters per hour |
| 5.080 | millimeters per minute |
| 0.084667 | millimeters per second |

## FOOT PER MINUTE: =

| | |
|---|---|
| 0.27273 | miles per day |
| 0.011364 | miles per hour |
| 0.0001893939 | miles per minute |
| 0.0000031566 | miles per second |
| 0.43891 | kilometers per day |
| 0.018288 | kilometers per hour |
| 0.0003048006 | kilometers per minute |
| 0.00000508 | kilometers per second |
| 2.18182 | furlongs per day |
| 0.090909 | furlongs per hour |
| 0.0015151 | furlongs per minute |
| 0.000025253 | furlongs per second |
| 4.38913 | hektometers per day |
| 0.18288 | hektometers per hour |
| 0.003048006 | hektometers per minute |
| 0.0000508 | hektometers per second |
| 21.81818 | chains per day |
| 0.90909 | chains per hour |
| 0.015151 | chains per minute |
| 0.00025253 | chains per second |
| 43.89129 | dekameters per day |
| 1.82880 | dekameters per hour |
| 0.03048006 | dekameters per minute |
| 0.000508 | dekameters per second |
| 87.27273 | rods per day |
| 3.63636 | rods per hour |
| 0.0606061 | rods per minute |
| 0.0010101 | rods per second |
| 438.91286 | meters per day |
| 18.28804 | meters per hour |

FOOT PER MINUTE (cont'd): =

| | |
|---|---|
| 0.3048006 | meters per minute |
| 0.00508 | meters per second |
| 480.0 | yards per day |
| 20.0 | yards per hour |
| 0.33333 | yards per minute |
| 0.0055556 | yards per second |
| 518.400 | varas (Texas) per day |
| 21.60 | varas (Texas) per hour |
| 0.3600 | varas (Texas) per minute |
| 0.0060000 | varas (Texas) per second |
| 1,440.0 | feet per day |
| 60.0 | feet per hour |
| 1.0 | feet per minute |
| 0.016667 | feet per second |
| 1,920 | spans per day |
| 80.0 | spans per hour |
| 1.33333 | spans per minute |
| 0.022222 | spans per second |
| 2,181.81818 | links per day |
| 90.90909 | links per hour |
| 1.51515 | links per minute |
| 0.025253 | links per second |
| 4,320 | hands per day |
| 180.0 | hands per hour |
| 3.0 | hands per minute |
| 0.050000 | hands per second |
| 4,389.12864 | decimeters per day |
| 182.88036 | decimeters per hour |
| 3.048006 | decimeters per minute |
| 0.050800 | decimeters per second |
| 43,891.2864 | centimeters per day |
| 1,828.80360 | centimeters per hour |
| 30.48006 | centimeters per minute |
| 0.50800 | centimeters per second |
| 17,280 | inches per day |
| 720 | inches per hour |
| 12.0 | inches per minute |
| 0.20000 | inches per second |
| 438,912.864 | millimeters per day |
| 18,288.0360 | millimeters per hour |
| 304.8006 | millimeters per minute |
| 5.08001 | millimeters per second |

FOOT PER SECOND: =

|  |  |
|---|---|
| 16.36363 | miles per day |
| 0.68182 | miles per hour |
| 0.011364 | miles per minute |
| 0.0001893939 | miles per second |
| 26.33477 | kilometers per day |
| 1.097282 | kilometers per hour |
| 0.018288 | kilometers per minute |
| 0.0003048006 | kilometers per second |
| 130.90909 | furlongs per day |
| 5.45455 | furlongs per hour |
| 0.090909 | furlongs per minute |
| 0.0015151 | furlongs per second |
| 263.34772 | hektometers per day |
| 10.97282 | hektometers per hour |
| 0.18288 | hektometers per minute |
| 0.003048006 | hektometers per second |
| 1,309.09090 | chains per day |
| 54.54545 | chains per hour |
| 0.90909 | chains per minute |
| 0.015151 | chains per second |
| 2,633.47718 | dekameters per day |
| 109.72823 | dekameters per hour |
| 1.82880 | dekameters per minute |
| 0.03048006 | dekameters per second |
| 5,236.36364 | rods per day |
| 218.18182 | rods per hour |
| 3.63636 | rods per minute |
| 0.0606061 | rods per second |
| 26,334.77184 | meters per day |
| 1,097.28216 | meters per hour |
| 18.28804 | meters per minute |
| 0.3048006 | meters per second |
| 28,800 | yards per day |
| 1,200 | yards per hour |
| 20.0 | yards per minute |
| 0.33333 | yards per second |
| 31,104 | varas (Texas) per day |
| 1,296.0 | varas (Texas) per hour |
| 21.60 | varas (Texas) per minute |
| 0.3600 | varas (Texas) per second |
| 86,400 | feet per day |
| 3,600 | feet per hour |
| 60 | feet per minute |
| 1 | feet per second |

**FOOT PER SECOND (cont'd): =**

115,200 . . . . . . . . . . . . . . . . . . . . . . . . . . . . . . . . . .spans per day
4,800 . . . . . . . . . . . . . . . . . . . . . . . . . . . . . . . . . spans per hour
80.0 . . . . . . . . . . . . . . . . . . . . . . . . . . . . . . . . .spans per minute
1.33333 . . . . . . . . . . . . . . . . . . . . . . . . . . . . . . . .spans per second
130,909.090909 . . . . . . . . . . . . . . . . . . . . . . . . . . links per day
5,454.54545 . . . . . . . . . . . . . . . . . . . . . . . . . . . . links per hour
90.90909 . . . . . . . . . . . . . . . . . . . . . . . . . . . . . . links per minute
1.51515 . . . . . . . . . . . . . . . . . . . . . . . . . . . . . . . links per second
259,200 . . . . . . . . . . . . . . . . . . . . . . . . . . . . . . . hands per day
10,800 . . . . . . . . . . . . . . . . . . . . . . . . . . . . . . . .hands per hour
180.0 . . . . . . . . . . . . . . . . . . . . . . . . . . . . . . . hands per minute
3.0 . . . . . . . . . . . . . . . . . . . . . . . . . . . . . . . . .hands per second
263,347.7184 . . . . . . . . . . . . . . . . . . . . . . . . . . .decimeters per day
10,972.8216 . . . . . . . . . . . . . . . . . . . . . . . . . . . . decimeters per hour
182.88036 . . . . . . . . . . . . . . . . . . . . . . . . . . . . .decimeters per minute
3.048006 . . . . . . . . . . . . . . . . . . . . . . . . . . . . . .decimeters per second
2,633,477.1840 . . . . . . . . . . . . . . . . . . . . . . . . centimeters per day
109,728.2160 . . . . . . . . . . . . . . . . . . . . . . . . . . centimeters per hour
1,828.8036 . . . . . . . . . . . . . . . . . . . . . . . . . . . . centimeters per minute
30.48006 . . . . . . . . . . . . . . . . . . . . . . . . . . . . . centimeters per second
1,036,800 . . . . . . . . . . . . . . . . . . . . . . . . . . . . . inches per day
43,200 . . . . . . . . . . . . . . . . . . . . . . . . . . . . . . . .inches per hour
720 . . . . . . . . . . . . . . . . . . . . . . . . . . . . . . . . . inches per minute
12 . . . . . . . . . . . . . . . . . . . . . . . . . . . . . . . . . . inches per second
26,334,771.8400 . . . . . . . . . . . . . . . . . . . . . . . millimeters per day
1,097,282.1600 . . . . . . . . . . . . . . . . . . . . . . . . .millimeters per hour
18,288.0360 . . . . . . . . . . . . . . . . . . . . . . . . . . . millimeters per minute
304.8006 . . . . . . . . . . . . . . . . . . . . . . . . . . . . . millimeters per second

**FOOT POUNDS: =**

32.174 . . . . . . . . . . . . . . . . . . . . . . . . . . . . . . . . foot poundals
386.088 . . . . . . . . . . . . . . . . . . . . . . . . . . . . . . . inch poundals
1.0 . . . . . . . . . . . . . . . . . . . . . . . . . . . . . . . . . . foot pounds
12 . . . . . . . . . . . . . . . . . . . . . . . . . . . . . . . . . . . inch pounds
0.00000035703 . . . . . . . . . . . . . . . . . . . . . . . . . ton (short) calories
0.00032389 . . . . . . . . . . . . . . . . . . . . . . . . . . . .kilogram calories
0.00071406 . . . . . . . . . . . . . . . . . . . . . . . . . . . . pound calories
0.0032389 . . . . . . . . . . . . . . . . . . . . . . . . . . . . .hektogram calories
0.032389 . . . . . . . . . . . . . . . . . . . . . . . . . . . . . dekagram calories
0.011425 . . . . . . . . . . . . . . . . . . . . . . . . . . . . . . ounce calories
0.32389 . . . . . . . . . . . . . . . . . . . . . . . . . . . . . . . gram calories
3.2389 . . . . . . . . . . . . . . . . . . . . . . . . . . . . . . .decigram calories
32.389 . . . . . . . . . . . . . . . . . . . . . . . . . . . . . . centigram calories

FOOT POUNDS (cont'd): =

| | |
|---|---|
| 323.89 | milligram calories |
| 0.0000000156925 | kilowatt days |
| 0.00000037662 | kilowatt hours |
| 0.0000225972 | kilowatt minutes |
| 0.00135583 | kilowatt seconds |
| 0.0000156925 | watt days |
| 0.00037662 | watt hours |
| 0.0225972 | watt minutes |
| 1.35583 | watt seconds |
| 0.0015240 | ton meters (short) |
| 0.138255 | kilogram meters |
| 0.30480 | pound meters |
| 1.38255 | hektogram meters |
| 13.8255 | dekagram meters |
| 4.8768 | ounce meters |
| 138.255 | gram meters |
| 1,382.55 | decigram meters |
| 13,825.5 | centigram meters |
| 138,255 | milligram meters |
| 0.000015240 | ton hektometers |
| 0.00138255 | kilogram hektometers |
| 0.0030480 | pound hektometers |
| 0.0138255 | hektogram hektometers |
| 0.138255 | dekagram hektometers |
| 0.048768 | ounce hektometers |
| 1.38255 | gram hektometers |
| 13.8255 | decigram hektometers |
| 138.255 | centigram hektometers |
| 1,382.55 | milligram hektometers |
| 0.00015240 | ton dekameters |
| 0.0138255 | kilogram dekameters |
| 0.030480 | pound dekameters |
| 0.138255 | hektogram dekameters |
| 1.38255 | dekagram dekameters |
| 0.48768 | ounce dekameters |
| 13.8255 | gram dekameters |
| 138.255 | decigram dekameters |
| 1,382.55 | centigram dekameters |
| 13,825.5 | milligram dekameters |
| 0.0050000 | ton feet (short) |
| 0.45359 | kilogram feet |
| 1.0 | pound feet |
| 4.53592 | hektogram feet |
| 45.35916 | dekagram feet |

FOOT POUNDS (cont'd): =

| | |
|---|---|
| 16 | ounce feet |
| 453.59157 | gram feet |
| 4,535.91566 | decigram feet |
| 45,359.15664 | centigram feet |
| 453,591.56642 | milligram feet |
| 0.060000 | ton inches |
| 5.44308 | kilogram inches |
| 12.0 | pound inches |
| 54.43104 | hektogram inches |
| 544.30992 | dekagram inches |
| 192 | ounce inches |
| 5,443.09984 | gram inches |
| 54,430.98792 | decigram inches |
| 544,309.87968 | centigram inches |
| 5,443,098.7968 | milligram inches |
| 0.015240 | ton decimeters |
| 1.38255 | kilogram decimeters |
| 3.0480 | pound decimeters |
| 13,8255 | hektogram decimeters |
| 138.255 | dekagram decimeters |
| 48.760 | ounce decimeters |
| 1,382.55 | gram decimeters |
| 13,825.5 | decigram decimeters |
| 138,255 | centigram decimeters |
| 1,382,550 | milligram decimeters |
| 0.15240 | ton centimeters |
| 13.8255 | kilogram centimeters |
| 30.480 | pound centimeters |
| 138.255 | hektogram centimeters |
| 1,382.55 | dekagram centimeters |
| 487.60 | ounce centimeters |
| 13,825.5 | gram centimeters |
| 138,255 | decigram centimeters |
| 1,382,550 | centigram centimeters |
| 13,825,500 | milligram centimeters |
| 1.5240 | ton millimeters |
| 138.255 | kilogram millimeters |
| 304.80 | pound millimeters |
| 1,382.55 | hektogram millimeters |
| 13,825.5 | dekagram millimeters |
| 4,876.0 | ounce millimeters |
| 138,255 | gram millimeters |
| 1,382,550 | decigram millimeters |
| 13,825,500 | centigram millimeters |

**FOOT POUNDS (cont'd): =**

| | |
|---|---|
| 138,255,000 | milligram millimeters |
| 0.000013381 | kiloliter-atmospheres |
| 0.00013381 | hektoliter-atmospheres |
| 0.0013381 | dekaliter-atmospherees |
| 0.013381 | liter-atmospheres |
| 0.13381 | deciliter-atmospheres |
| 1.3381 | centiliter-atmospheres |
| 13.381 | milliliter-atmospheres |
| 0.0000000000000013381 | cubic kilometer-atmospheres |
| 0.000000000013381 | cubic hektometer-atmospheres |
| 0.000000013381 | cubic dekameter-atmospheres |
| 0.000013381 | cubic meter-atmospheres |
| 0.00047253 | cubic feet-atmospheres |
| 0.013381 | cubic decimeter-atmospheres |
| 13.381 | cubic centimeter-atmospheres |
| 13,381 | cubic millimeter-atmospheres |
| 1.35582 | joules (absolute) |
| 0.00000049814 | Cheval-Vapeur hours |
| 0.0000000210438 | horsepower day |
| 0.00000050505 | horsepower hours |
| 0.0000303030 | horsepower minutes |
| 0.00181818 | horsepower seconds |
| 0.0000008808 | pounds of carbon oxidized with 100% efficiency |
| 0.0000013256 | pounds of water evaporated from and at 212° F. |
| 0.0012854 | B.T.U. |
| 13,558,200 | ergs |
| 13,558,200 | centimeters dynes |

**FOOT POUND PER DAY: =**

| | |
|---|---|
| 32.174 | foot poundals per day |
| 1.34058 | foot poundals per hour |
| 0.022343 | foot poundals per minute |
| 0.00037238 | foot poundals per second |
| 1.0 | foot pounds per day |
| 0.041667 | foot pounds per hour |
| 0.00069444 | foot pounds per minute |
| 0.000011574 | foot pounds per second |
| 0.00032389 | kilogram calories per day |
| 0.000013495 | kilogram calories per hour |
| 0.00000022492 | kilogram calories per minute |

## FOOT POUND PER DAY (cont'd): =

| | |
|---|---|
| 0.0000000037487 | kilogram calories per second |
| 0.011425 | ounce calories per day |
| 0.00047604 | ounce calories per hour |
| 0.0000079340 | ounce calories per minute |
| 0.00000013223 | ounce calories per second |
| 0.138255 | kilogram meters per day |
| 0.0057606 | kilogram meters per hour |
| 0.00009601 | kilogram meters per minute |
| 0.0000016002 | kilogram meters per second |
| 0.013381 | liter-atmospheres per day |
| 0.00055754 | liter-atmospheres per hour |
| 0.0000092924 | liter-atmospheres per minute |
| 0.00000015487 | liter-atmospheres per second |
| 0.00047253 | cubic foot atmospheres per day |
| 0.000019689 | cubic foot atmospheres per hour |
| 0.00000032815 | cubic foot atmospheres per minute |
| 0.0000000054691 | cubic foot atmospheres per second |
| 0.000000021333 | Cheval-Vapeurs |
| 0.000000021044 | horsepower |
| 0.000000015692 | kilowatts |
| 0.000015692 | watts |
| 1.35582 | joules per day |
| 0.056492 | joules per hour |
| 0.00094154 | joules per minute |
| 0.000015692 | joules per second |
| 0.00000008808 | pounds of carbon oxidized with perfect efficiency per day |
| 0.0000000036700 | pounds of carbon oxidized with perfect efficiency per hour |
| 0.0000000000611667 | pounds of carbon oxidized with perfect efficiency per minute |
| 0.0000000000010194 | pounds of carbon oxidized with perfect efficiency per second |
| 0.0000013256 | pounds of water evaporated from and at 212° F. per day |
| 0.000000055233 | pounds of water evaporated from and at 212° F. per hour |
| 0.00000000092056 | pounds of water evaporated from and at 212° F. per minute |
| 0.000000000015343 | pounds of water evaporated from and at 212° F. per second |
| 0.0012854 | BTU per day |
| 0.000053558 | BTU per hour |

FOOT POUND PER DAY (cont'd): =

| | |
|---|---|
| 0.00000089264 | BTU per minute |
| 0.000000014877 | BTU per second |

FOOT POUND PER HOUR: =

| | |
|---|---|
| 722.1760 | foot poundals per day |
| 32.174 | foot poundals per hour |
| 0.53623 | foot poundals per minute |
| 0,0089372 | foot poundals per second |
| 24.0 | foot pounds per day |
| 1.0 | foot pounds per hour |
| 0.016667 | foot pounds per minute |
| 0.00027778 | foot pounds per second |
| 0.0077734 | kilogram calories per day |
| 0.00032389 | kilogram calories per hour |
| 0.0000053982 | kilogram calories per minute |
| 0.000000089969 | kilogram calories per second |
| 0.2742 | ounce calories per day |
| 0.011425 | ounce calories per hour |
| 0.00019042 | ounce calories per minute |
| 0.0000031736 | ounce calories per second |
| 3.31812 | kilogram meters per day |
| 0.138255 | kilogram meters per hour |
| 0.0023042 | kilogram meters per minute |
| 0.000038404 | kilogram meters per second |
| 0.32114 | liter-atmospheres per day |
| 0.013381 | liter-atmospheres per hour |
| 0.00022302 | liter-atmospheres per minute |
| 0.0000037169 | liter-atmospheres per second |
| 0.011341 | cubic foot atmospheres per day |
| 0.00047253 | cubic foot atmospheres per hour |
| 0.0000078755 | cubic foot atmospheres per minute |
| 0.00000013126 | cubic foot atmospheres per second |
| 0.0000005120 | Cheval-Vapeurs |
| 0.00000050505 | horsepower |
| 0.00000037661 | kilowatts |
| 0.00037661 | watts |
| 32.53968 | joules per day |
| 1.35582 | joules per hour |
| 0.022597 | joules per minute |
| 0.00037662 | joules per second |
| 0.0000021139 | pounds of carbon oxidized with perfect efficiency per day |
| 0.00000008808 | pounds of carbon oxidized with perfect efficiency per hour |

## FOOT POUND PER HOUR (cont'd): =

0.0000000014680 . . . . . . . . . . . . . . . . . . . . pounds of carbon oxidized with perfect efficiency per minute
0.000000000024467 . . . . . . . . . . . . . . pounds of carbon oxidized with perfect efficiency per second
0.000031814 . . . . . . . . . . . . . . . . . . . . pounds of water evaporated from and at 212° F. per day
0.0000013256 . . . . . . . . . . . . . . . . . . . pounds of water evaporated from and at 212° F. per hour
0.000000022093 . . . . . . . . . . . . . . . . . pounds of water evaporated from and at 212° F. per minute
0.00000000036822 . . . . . . . . . . . . . . . pounds of water evaporated from and at 212° F. per second
0.030850 . . . . . . . . . . . . . . . . . . . . . . . . . . . . . . . . . BTU per day
0.0012854 . . . . . . . . . . . . . . . . . . . . . . . . . . . . . . . BTU per hour
0.000021423 . . . . . . . . . . . . . . . . . . . . . . . . . . . . BTU per minute
0.00000035706 . . . . . . . . . . . . . . . . . . . . . . . . . . BTU per second

## FOOT POUND PER MINUTE: =

46,330.56 . . . . . . . . . . . . . . . . . . . . . . . . . . . . .foot poundals per day
1,930.44 . . . . . . . . . . . . . . . . . . . . . . . . . . . . . . foot poundals per hour
32.174 . . . . . . . . . . . . . . . . . . . . . . . . . . . . . .foot poundals per minute
0.53623 . . . . . . . . . . . . . . . . . . . . . . . . . . . . . foot poundals per second
1,440 . . . . . . . . . . . . . . . . . . . . . . . . . . . . . . . . .foot pounds per day
60 . . . . . . . . . . . . . . . . . . . . . . . . . . . . . . . . . . . foot pounds per hour
1.0 . . . . . . . . . . . . . . . . . . . . . . . . . . . . . . . . . .foot pounds per minute
0.016667 . . . . . . . . . . . . . . . . . . . . . . . . . . . . . . foot pounds per second
0.46640 . . . . . . . . . . . . . . . . . . . . . . . . . . . . . kilogram calories per day
0.019433 . . . . . . . . . . . . . . . . . . . . . . . . . . . kilogram calories per hour
0.00032389 . . . . . . . . . . . . . . . . . . . . . . . . kilogram calories per minute
0.0000053982 . . . . . . . . . . . . . . . . . . . . . . . kilogram calories per second
16.452 . . . . . . . . . . . . . . . . . . . . . . . . . . . . . . .ounce calories per day
0.6855 . . . . . . . . . . . . . . . . . . . . . . . . . . . . . . . ounce calories per hour
0.011425 . . . . . . . . . . . . . . . . . . . . . . . . . . . . .ounce calories per minute
0.00019042 . . . . . . . . . . . . . . . . . . . . . . . . . . .ounce calories per second
199.0872 . . . . . . . . . . . . . . . . . . . . . . . . . . . . . kilogram meters per day
8.29530 . . . . . . . . . . . . . . . . . . . . . . . . . . . . . kilogram meters per hour
0.138255 . . . . . . . . . . . . . . . . . . . . . . . . . . kilogram meters per minute
0.0023043 . . . . . . . . . . . . . . . . . . . . . . . . . .kilogram meters per second
19.26864 . . . . . . . . . . . . . . . . . . . . . . . . . . . . liter-atmospheres per day
0.80286 . . . . . . . . . . . . . . . . . . . . . . . . . . . . liter-atmospheres per hour
0.013381 . . . . . . . . . . . . . . . . . . . . . . . . . .liter-atmospheres per minute

FOOT POUND PER MINUTE (cont'd): =

| | |
|---|---|
| 0.00022302 | liter-atmospheres per second |
| 0.68044 | cubic foot atmospheres per day |
| 0.028352 | cubic foot atmospheres per hour |
| 0.00047253 | cubic foot atmospheres per minute |
| 0.0000078755 | cubic foot atmospheres per second |
| 0.000030719 | Cheval-Vapeurs |
| 0.000030303 | horsepower |
| 0.000022597 | kilowatts |
| 0.022597 | watts |
| 1,952.3808 | joules per day |
| 81.34920 | joules per hour |
| 1.35582 | joules per minute |
| 0.022597 | joules per second |
| 0.00012684 | pounds of carbon oxidized with perfect efficiency per day |
| 0.0000052848 | pounds of carbon oxidized with perfect efficiency per hour |
| 0.00000008808 | pounds of carbon oxidized with perfect efficiency per minute |
| 0.000000001468 | pounds of carbon oxidized with perfect efficiency per second |
| 0.0019089 | pounds of water evaporated from and at 212° F. per day |
| 0.000079536 | pounds of water evaporated from and at 212° F. per hour |
| 0.0000013256 | pounds of water evaporated from and at 212° F. per minute |
| 0.000000022093 | pounds of water evaporated from and at 212° F. per second |
| 1.85098 | BTU per day |
| 0.077124 | BTU per hour |
| 0.0012854 | BTU per minute |
| 0.000021423 | BTU per second |

FOOT POUND PER SECOND: =

| | |
|---|---|
| 2,779,833.6 | foot poundals per day |
| 115,826.40 | foot poundals per hour |
| 1,930.440 | foot poundals per minute |
| 32.174 | foot poundals per second |
| 86,400 | foot pounds per day |
| 3,600 | foot pounds per hour |
| 60 | foot pounds per minute |

## FOOT POUND PER SECOND (cont'd): =

1.0 . . . . . . . . . . . . . . . . . . . . . . . . . . . . . foot pounds per second
27.9841 . . . . . . . . . . . . . . . . . . . . . . . . . kilogram calories per day
1.1660 . . . . . . . . . . . . . . . . . . . . . . . . . kilogram calories per hour
0.019433 . . . . . . . . . . . . . . . . . . . . . . kilogram calories per minute
0.00032389 . . . . . . . . . . . . . . . . . . . . . kilogram calories per second
987.120 . . . . . . . . . . . . . . . . . . . . . . . . . . .ounce calories per day
41.130 . . . . . . . . . . . . . . . . . . . . . . . . . . . ounce calories per hour
0.68550 . . . . . . . . . . . . . . . . . . . . . . . . .ounce calories per minute
0.011425 . . . . . . . . . . . . . . . . . . . . . . . .ounce calories per second
11,945.2320 . . . . . . . . . . . . . . . . . . . . . kilogram meters per day
497.7180 . . . . . . . . . . . . . . . . . . . . . . . kilogram meters per hour
8.29530 . . . . . . . . . . . . . . . . . . . . . . . kilogram meters per minute
0.138255 . . . . . . . . . . . . . . . . . . . . . . .kilogram meters per second
1,156.11840 . . . . . . . . . . . . . . . . . . . . . liter-atmospheres per day
48.17160 . . . . . . . . . . . . . . . . . . . . . . . liter-atmospheres per hour
0.80286 . . . . . . . . . . . . . . . . . . . . . . .liter-atmospheres per minute
0.013381 . . . . . . . . . . . . . . . . . . . . . . liter-atmospheres per second
40.82659 . . . . . . . . . . . . . . . . . . . . . cubic foot atmospheres per day
1.70111 . . . . . . . . . . . . . . . . . . . . . .cubic foot atmospheres per hour
0.028352 . . . . . . . . . . . . . . . . . . . cubic foot atmoshperes per minute
0.00047253 . . . . . . . . . . . . . . . . . . cubic foot atmospheres per second
0.0018432 . . . . . . . . . . . . . . . . . . . . . . . . . . . . . . .Cheval-Vapeurs
0.0018182 . . . . . . . . . . . . . . . . . . . . . . . . . . . . . . . . . .horsepower
0.0013558 . . . . . . . . . . . . . . . . . . . . . . . . . . . . . . . . . . . kilowatts
1.35582 . . . . . . . . . . . . . . . . . . . . . . . . . . . . . . . . . . . . . . . watts
117,142.8480 . . . . . . . . . . . . . . . . . . . . . . . . . . . . . joules per day
4,880.9520 . . . . . . . . . . . . . . . . . . . . . . . . . . . . . . . joules per hour
81.34920 . . . . . . . . . . . . . . . . . . . . . . . . . . . . . . joules per minute
1.35582 . . . . . . . . . . . . . . . . . . . . . . . . . . . . . . . .joules per second
0.0076101 . . . . . . . . . . . . . . . . . . . . . pounds of carbon oxidized with perfect efficiency per day
0.00031709 . . . . . . . . . . . . . . . . . . . . pounds of carbon oxidized with perfect efficiency per hour
0.0000052848 . . . . . . . . . . . . . . . . . . pounds of carbon oxidized with perfect efficiency per minute
0.00000008808 . . . . . . . . . . . . . . . . . pounds of carbon oxidized with perfect efficiency per second
0.11453 . . . . . . . . . . . . . . . . . . . . . . pounds of water evaporated from and at 212° F. per day
0.0047722 . . . . . . . . . . . . . . . . . . .pounds of water evaporated from and at 212° F. per hour
0.000079536 . . . . . . . . . . . . . . . . . . . pounds of water evaporated from and at 212° F. per minute

FOOT POUND PER SECOND (cont'd): =

0.0000013256 . . . . . . . . . . . . . . . . . . . . pounds of water evaporated from and 212° F. per second
111.05856 . . . . . . . . . . . . . . . . . . . . . . . . . . . . . . BTU per day
4.62744 . . . . . . . . . . . . . . . . . . . . . . . . . . . . . . BTU per hour
0.077124 . . . . . . . . . . . . . . . . . . . . . . . . . . . . . BTU per minute
0.0012854 . . . . . . . . . . . . . . . . . . . . . . . . . . . . BTU per second

FOOT OF WATER AT 60° F: =

0.00304801 . . . . . . . . . . . . . . . . . . . . hektometers of water at 60° F.
0.0304801 . . . . . . . . . . . . . . . . . . . . . . dekameters of water at 60° F.
0.304801 . . . . . . . . . . . . . . . . . . . . . . . . . meters of water at 60° F.
1.0 . . . . . . . . . . . . . . . . . . . . . . . . . . . . . feet of water at 60° F.
12 . . . . . . . . . . . . . . . . . . . . . . . . . . . . inches of water at 60° F.
3.04801 . . . . . . . . . . . . . . . . . . . . . . . decimeters of water at 60° F.
30.4801 . . . . . . . . . . . . . . . . . . . . . . centimeters of water at 60° F.
304.801 . . . . . . . . . . . . . . . . . . . . . . . millimeters of water at 60° F.
0.002240 . . . . . . . . . . . . . . . . . . . . hektometers of mercury at 32° F.
0.02240 . . . . . . . . . . . . . . . . . . . . . . dekameters of mercury at 32° F.
0.2240 . . . . . . . . . . . . . . . . . . . . . . . . . . meters of mercury at 32° F.
0.73491 . . . . . . . . . . . . . . . . . . . . . . . . . . feet of mercury at 32° F.
8.81888 . . . . . . . . . . . . . . . . . . . . . . . . ounces of mercury at 32° F.
0.882612 . . . . . . . . . . . . . . . . . . . . . . . . inches of mercury at 32° F.
2.240 . . . . . . . . . . . . . . . . . . . . . . . . decimeters of mercury at 32° F.
22.40 . . . . . . . . . . . . . . . . . . . . . . . centimeters of mercury at 32° F.
224.0 . . . . . . . . . . . . . . . . . . . . . . . millimeters of mercury at 32° F.
3,359.85556 . . . . . . . . . . . . . . . . . . . . . . . . tons per square hektometer
33.59856 . . . . . . . . . . . . . . . . . . . . . . . . . . tons per square dekameter
0.33599 . . . . . . . . . . . . . . . . . . . . . . . . . . . . . tons per square meter
0.031214 . . . . . . . . . . . . . . . . . . . . . . . . . . . . . tons per square foot
0.00021677 . . . . . . . . . . . . . . . . . . . . . . . . . . . tons per square inch
0.0033599 . . . . . . . . . . . . . . . . . . . . . . . . . . tons per square decimeter
0.000033599 . . . . . . . . . . . . . . . . . . . . . . . . tons per square centimeter
0.00000033599 . . . . . . . . . . . . . . . . . . . . . . . tons per square millimeter
3,048,010 . . . . . . . . . . . . . . . . . . . . . . kilograms per square hektometer
30,480.1 . . . . . . . . . . . . . . . . . . . . . . . kilograms per square dekameter
304.801 . . . . . . . . . . . . . . . . . . . . . . . . . . kilograms per square meter
28.31705 . . . . . . . . . . . . . . . . . . . . . . . . . . kilograms per square foot
0.19665 . . . . . . . . . . . . . . . . . . . . . . . . . . . kilograms per square inch
3.04801 . . . . . . . . . . . . . . . . . . . . . . . . kilograms per square decimeter
0.0304801 . . . . . . . . . . . . . . . . . . . . . . kilograms per square centimeter
0.000304801 . . . . . . . . . . . . . . . . . . . . . kilograms per square millimeter
6,719,721.9 . . . . . . . . . . . . . . . . . . . . . . . . pounds per square hektometer

FOOT OF WATER AT 60° F. (cont.d): =

67,197.2 . . . . . . . . . . . . . . . . . . . . . . . . pounds per square dekameter
671.97219 . . . . . . . . . . . . . . . . . . . . . . . . pounds per square meter
62.42832 . . . . . . . . . . . . . . . . . . . . . . . . pounds per square foot
0.433530 . . . . . . . . . . . . . . . . . . . . . . . . pounds per square inch
6.71972 . . . . . . . . . . . . . . . . . . . . . . . . pounds per square decimeter
0.067197 . . . . . . . . . . . . . . . . . . . . . . . . pounds per square centimeter
0.00067197 . . . . . . . . . . . . . . . . . . . . . . . . pounds per square millimeter
30,480,100 . . . . . . . . . . . . . . . . . . . . . . . . hektograms per square hektometer
304,801 . . . . . . . . . . . . . . . . . . . . . . . . hektograms per square dekameter
3,048.01 . . . . . . . . . . . . . . . . . . . . . . . . hektograms per square meter
283.1705 . . . . . . . . . . . . . . . . . . . . . . . . hektograms per square foot
1.9665 . . . . . . . . . . . . . . . . . . . . . . . . hektograms per square inch
30.4801 . . . . . . . . . . . . . . . . . . . . . . . . hektograms per square decimeter
0.304801 . . . . . . . . . . . . . . . . . . . . . . . . hektograms per square centimeter
0.00304801 . . . . . . . . . . . . . . . . . . . . . . . . hektograms per square millimeter
304,801,000 . . . . . . . . . . . . . . . . . . . . . . . . dekagrams per square hektometer
3,048,010 . . . . . . . . . . . . . . . . . . . . . . . . dekagrams per square dekameter
30,480.1 . . . . . . . . . . . . . . . . . . . . . . . . dekagrams per square meter
2,831.705 . . . . . . . . . . . . . . . . . . . . . . . . dekagrams per square foot
19.665 . . . . . . . . . . . . . . . . . . . . . . . . dekagrams per square inch
304.801 . . . . . . . . . . . . . . . . . . . . . . . . dekagrams per square decimeter
3.04801 . . . . . . . . . . . . . . . . . . . . . . . . dekagrams per square centimeter
0.0304801 . . . . . . . . . . . . . . . . . . . . . . . . dekagrams per square millimeter
107,515,550.4 . . . . . . . . . . . . . . . . . . . . . . . . ounces per square hektometer
1,075,155.504 . . . . . . . . . . . . . . . . . . . . . . . . ounces per square dekameter
10,751.55504 . . . . . . . . . . . . . . . . . . . . . . . . ounces per square meter
998.85312 . . . . . . . . . . . . . . . . . . . . . . . . ounces per square foot
6.93648 . . . . . . . . . . . . . . . . . . . . . . . . ounces per square inch
107.51556 . . . . . . . . . . . . . . . . . . . . . . . . ounces per square decimeter
1.075156 . . . . . . . . . . . . . . . . . . . . . . . . ounces per square centimeter
0.0107516 . . . . . . . . . . . . . . . . . . . . . . . . ounces per square millimeter
3,048,010,000 . . . . . . . . . . . . . . . . . . . . . . . . grams per square hektometer
30,480,100 . . . . . . . . . . . . . . . . . . . . . . . . grams per square dekameter
304,801 . . . . . . . . . . . . . . . . . . . . . . . . grams per square meter
28,317.05 . . . . . . . . . . . . . . . . . . . . . . . . grams per square foot
196.65 . . . . . . . . . . . . . . . . . . . . . . . . grams per square inch
3,048.01 . . . . . . . . . . . . . . . . . . . . . . . . grams per square decimeter
30.4801 . . . . . . . . . . . . . . . . . . . . . . . . grams per square centimeter
0.304801 . . . . . . . . . . . . . . . . . . . . . . . . grams per square millimeter
30,480,100,000 . . . . . . . . . . . . . . . . . . . . . . . . decigrams per square hektometer
304,801,000 . . . . . . . . . . . . . . . . . . . . . . . . decigrams per square dekameter
3,048,010 . . . . . . . . . . . . . . . . . . . . . . . . decigrams per square meter
283,170.5 . . . . . . . . . . . . . . . . . . . . . . . . decigrams per square foot
1,966.5 . . . . . . . . . . . . . . . . . . . . . . . . decigrams per square inch

FOOT OF WATER AT 60° F. (cont'd): =

| | |
|---|---|
| 30,480.1 | decigrams per square decimeter |
| 304.801 | decigrams per square centimeter |
| 3.04801 | decigrams per square millimeter |
| 304,801,000,000 | centigrams per square hektometer |
| 3,048,010,000 | centigrams per square dekameter |
| 30,480,100 | centigrams per square meter |
| 2,831,705 | centigrams per square foot |
| 19,665 | centigrams per square inch |
| 304,801 | centigrams per square decimeter |
| 3,048.01 | centigrams per square centimeter |
| 30.4801 | centigrams per square millimeter |
| 3,048,010,000,000 | milligrams per square hektometer |
| 30,480,100,000 | milligrams per square dekameter |
| 304,801,000 | milligrams per square meter |
| 28,317,050 | milligrams per square foot |
| 196,650 | milligrams per square inch |
| 3,048,010 | milligrams per square decimeter |
| 30,480.1 | milligrams per square centimeter |
| 304.801 | milligrams per square millimeter |
| 0.029889 | bars |
| 0.0294979 | atmospheres |
| 821.2 | feet of air @ 62° F. and 29.92 barometer pressure |
| 2,988,874,717,500 | dynes per square hektometer |
| 29,888,747,175 | dynes per square dekameter |
| 298,887,471.75 | dynes per square meter |
| 27,767,658.99497 | dynes per square foot |
| 192,830.60410 | dynes per square inch |
| 2,988,874.7175 | dynes per square decimeter |
| 29,888.74718 | dynes per square centimeter |
| 298.88747 | dynes per square millimeter |

GALLON (DRY): =

| | |
|---|---|
| 0.00044040 | kiloliters |
| 0.00044040 | cubic meters |
| 0.0057601 | cubic yards |
| 0.027701 | barrels |
| 0.0044040 | hektoliters |
| 0.12497 | bushels—U.S. (dry) |
| 0.12116 | bushels—Imperial (dry) |
| 0.15553 | cubic feet |
| 0.44040 | dekaliters |

GALLON (DRY) (cont'd): =

| | |
|---|---|
| 1.16342 | gallons—U.S. (liquid) |
| 1 | gallons—U.S. (dry) |
| 0.96874 | gallons—Imperial |
| 4.65368 | quarts (liquid) |
| 4 | quarts (dry) |
| 4.4040 | liters |
| 4.4040 | cubic decimeters |
| 9.30736 | pints (fluid) |
| 37.22943 | gills (fluid) |
| 44.04010 | deciliters |
| 268.75 | cubic inches |
| 440.40097 | centiliters |
| 4,404.00974 | milliliters |
| 4,404.00974 | cubic centimeters |
| 4.404,009.74 | cubic millimeters |
| 0.14139 | sacks of cement |
| 71,481 | minims |
| 148.91775 | ounces (fluid) |
| 1,191.34199 | drams (fluid) |
| 0.0043301 | tons (long) of water @ 62° F. |
| 0.0043996 | tons (metric) of water @ 62° F. |
| 0.0048498 | tons (short) of water @ 62° F. |
| 4.40005 | kilograms of water @ 62° F. |
| 9.69943 | pounds (Avoir) of water @ 62° F. |
| 11.78750 | pounds (Troy) of water @ 62° F. |
| 44.00054111 | hektograms |
| 440.0054111 | dekagrams |
| 141.45004 | ounces (Troy) of water @ 62° F. |
| 155.19091 | ounces (Avoir.) of water @ 62° F. |
| 1,131.60029 | drams (Troy) of water @ 62° F. |
| 2,483.05454 | drams (Avoir.) of water @ 62° F. |
| 2,829.00071 | pennyweights of water @ 62° F. |
| 3,394.80086 | scruples (Avoir.) of water @ 62° F. |
| 4,400.054111 | grams of water @ 62° F. |
| 44,000.54111 | decigrams of water at 62° F. |
| 67,896.022703 | grains of water at 62° F. |
| 440,005.41110 | centigrams of water at 62° F. |
| 4,400,054.1197 | milligrams of water at 62° F. |

GALLON (IMPERIAL): =

| | |
|---|---|
| 0.00045460 | kiloliters |
| 0.00045460 | cubic meters |

GALLON (IMPERIAL) (cont'd): =

| | |
|---|---|
| 0.0059459 | cubic yards |
| 0.028594 | barrels |
| 0.045460 | hektoliters |
| 0.12900 | bushels—U.S. (dry) |
| 0.125066 | bushels—Imperial (dry) |
| 0.16054 | cubic feet |
| 0.45460 | dekaliters |
| 1.20094 | gallons—U.S. (liquid) |
| 1.032184 | gallons—U.S. (dry) |
| 1 | gallon—Imperial |
| 4.80376 | quarts (liquid) |
| 4.12820 | quarts (dry) |
| 4.54596 | liters |
| 4.54596 | cubic decimeters |
| 9.60752 | pints (liquid) |
| 38.43008 | gills (liquid) |
| 45.4596 | deciliters |
| 277.41714 | cubic inches |
| 454.596 | centiliters |
| 4,545.96 | milliliters |
| 4,545.96 | cubic centimeters |
| 4,545,960 | cubic millimeters |
| 0.14595 | sacks of cement |
| 73,785.7536 | minims |
| 153.72032 | ounces (fluid) |
| 1,299.76256 | drams (fluid) |
| 0.0044698 | tons (long) of water @ 62° F. |
| 0.0045415 | tons (metric) of water @ 62° F. |
| 0.0050061 | tons (short) of water @ 62° F. |
| 4.54196 | kilograms of water @ 62° F. |
| 10.012237 | pounds (Avoir.) of water @ 62° F. |
| 12.16765 | pounds (Troy) of water @ 62° F. |
| 45.41955 | hektograms of water @ 62° F. |
| 454.19551 | dekagrams of water @ 62° F. |
| 146.011774 | ounces (Troy) of water @ 62° F. |
| 160.19579 | ounces (Avoir.) of water @ 62° F. |
| 1,168.094195 | drams (Troy) of water @ 62° F. |
| 2,563.13262 | drams (Avoir.) of water @ 62° F. |
| 2,902.23549 | pennyweights of water @ 62° F. |
| 3,504.28258 | scruples of water @ 62° F. |
| 4,541.95508 | grams of water @ 62° F. |
| 45,419.5508 | decigrams of water @ 62° F. |
| 70,085.65746 | grains of water @ 62° F. |

GALLON (IMPERIAL) (cont'd): =

| | |
|---|---|
| 454,195.508 | centigrams of water @ 62° F. |
| 4,541,955.08 | milligrams of water @ 62° F. |

GALLON (LIQUID): =

| | |
|---|---|
| 0.0037854 | kiloliters |
| 0.0037854 | cubic meters |
| 0.004951 | cubic yards |
| 0.0238095 | barrels |
| 0.037854 | hektoliters |
| 0.10742 | bushels—U.S. (dry) |
| 0.10414 | bushels—Imperial (dry) |
| 0.133681 | cubic feet |
| 0.37854 | dekaliters |
| 1 | gallons—U.S. (liquid) |
| 0.85948 | gallons—U.S. (dry) |
| 0.83268 | gallons—Imperial |
| 4 | quarts (liquid) |
| 3.43747 | quarts (dry) |
| 3.78544 | liters |
| 3.78544 | cubic decimeters |
| 8 | pints (liquid) |
| 32 | gills (liquid) |
| 37.8544 | deciliters |
| 231 | cubic inches |
| 378.544 | centiliters |
| 3,785.44 | milliliters |
| 3,785.44 | cubic centimeters |
| 3,785,440 | cubic millimeters |
| 0.12153 | sacks of cement |
| 61,440 | minims |
| 128 | ounces (fluid) |
| 1,024 | drams (fluid) |
| 0.0037219 | tons (long) of water @ 62° F. |
| 0.0037816 | tons (metric) of water @ 62° F. |
| 0.0041685 | tons (short) of water @ 62° F. |
| 3.7820 | kilograms of water @ 62° F. |
| 8.337 | pounds (Avoir.) of water @ 62° F. |
| 10.13177 | pounds (Troy) of water @ 62° F. |
| 37.820 | hektograms of water @ 62° F. |
| 378.20 | dekagrams of water @ 62° F. |
| 121.58124 | ounces (Troy) of water @ 62° F. |
| 133.392 | ounces (Avoir.) of water @ 62° F. |

GALLON (LIQUID) (cont'd): =

| | |
|---|---|
| 972.64992 | drams (Troy) of water @ 62° F. |
| 2,134.272 | drams (Avoir.) of water @ 62° F. |
| 2,431.6248 | pennyweights of water @ 62° F. |
| 2,917.94976 | scruples (Avoir.) of water @ 62° F. |
| 3,782 | grams of water @ 62° F. |
| 37,820 | decigrams of water @ 62° F. |
| 58,359 | grains of water @ 62° F. |
| 378,200 | centigrams of water @ 62° F. |
| 3,782,000 | milligrams of water @ 62° F. |

GALLON (U.S.) PER DAY: =

| | |
|---|---|
| 0.00037854 | kiloliters per day |
| 0.000015773 | kiloliters per hour |
| 0.00000026288 | kiloliters per minute |
| 0.0000000043813 | kiloliters per second |
| 0.00037854 | cubic meters per day |
| 0.000015773 | cubic meters per hour |
| 0.00000026288 | cubic meters per minute |
| 0.0000000043813 | cubic meters per second |
| 0.004951 | cubic yards per day |
| 0.00020629 | cubic yards per hour |
| 0.0000034382 | cubic yards per minute |
| 0.000000057303 | cubic yards per second |
| 0.02381 | barrels per day |
| 0.00099208 | barrels per hour |
| 0.000016535 | barrels per minute |
| 0.00000027558 | barrels per second |
| 0.0037854 | hektoliters per day |
| 0.00015773 | hektoliters per hour |
| 0.0000026288 | hektoliters per minute |
| 0.000000043813 | hektoliters per second |
| 0.10742 | bushels—U.S. (dry) per day |
| 0.0044758 | bushels—U.S. (dry) per hour |
| 0.000074597 | bushels—U.S. (dry) per minute |
| 0.0000012433 | bushels—U.S. (dry) per second |
| 0.10414 | bushels (Imperial) per day |
| 0.00433917 | bushels (Imperial) per hour |
| 0.0000723194 | bushels (Imperial) per minute |
| 0.00000120532 | bushels (Imperial) per second |
| 0.133681 | cubic feet per day |
| 0.00557002 | cubic feet per hour |
| 0.0000928337 | cubic feet per minute |

## GALLON (U.S.) PER DAY (cont'd): =

| | |
|---|---|
| 0.00000154723 | cubic feet per second |
| 0.378544 | dekaliters per day |
| 0.0157725 | dekaliters per hour |
| 0.000262875 | dekaliters per minute |
| 0.00000438128 | dekaliters per second |
| 1.0 | gallons—U.S. (liquid) per day |
| 0.0416667 | gallons—U.S. (liquid) per hour |
| 0.000694444 | gallons—U.S. (liquid) per minute |
| 0.0000115741 | gallons—U.S. (liquid) per second |
| 0.83268 | gallons (Imperial) per day |
| 0.0346950 | gallons (Imperial) per hour |
| 0.000578250 | gallons (Imperial) per minute |
| 0.00000963750 | gallons (Imperial) per second |
| 3.78544 | liters per day |
| 0.157726 | liters per hour |
| 0.00262877 | liters per minute |
| 0.0000438128 | liters per second |
| 3.78543 | cubic decimeters per day |
| 0.157726 | cubic decimeters per hour |
| 0.00262877 | cubic decimeters per minute |
| 0.0000438128 | cubic decimeters per second |
| 4.0 | quarts per day |
| 0.166667 | quarts per hour |
| 0.00277778 | quarts per minute |
| 0.0000462963 | quarts per second |
| 8.0 | pints per day |
| 0.333333 | pints per hour |
| 0.00555556 | pints per minute |
| 0.0000925925 | pints per second |
| 32.0 | gills per day |
| 1.333333 | gills per hour |
| 0.0222222 | gills per minute |
| 0.000370370 | gills per second |
| 37.8544 | deciliters per day |
| 1.57726 | deciliters per hour |
| 0.0262877 | deciliters per minute |
| 0.000438128 | deciliters per second |
| 231.0 | cubic inches per day |
| 9.625 | cubic inches per hour |
| 0.160417 | cubic inches per minute |
| 0.00267361 | cubic inches per second |
| 378.544 | centiliters per day |
| 15.772608 | centiliters per hour |
| 0.262877 | centiliters per minute |

**GALLON (U.S.) PER DAY (cont'd): =**

0.00438128 . . . . . . . . . . . . . . . . . . . . . . . . . centiliters per second
3,785.44 . . . . . . . . . . . . . . . . . . . . . . . . . . . . milliliters per day
157.726080 . . . . . . . . . . . . . . . . . . . . . . . . . . milliliters per hour
2.628768 . . . . . . . . . . . . . . . . . . . . . . . . . . . milliliters per minute
0.0438128 . . . . . . . . . . . . . . . . . . . . . . . . . . milliliters per second
3,785.44 . . . . . . . . . . . . . . . . . . . . . . . . . cubic centimeters per day
157.726080 . . . . . . . . . . . . . . . . . . . . . . cubic centimeters per hour
2.628768 . . . . . . . . . . . . . . . . . . . . . . . cubic centimeters per minute
0.0438128 . . . . . . . . . . . . . . . . . . . . . . cubic centimeters per second
3,785,440 . . . . . . . . . . . . . . . . . . . . . . . . cubic millimeters per day
157,726.249920 . . . . . . . . . . . . . . . . . . . cubic millimeters per hour
2,628.770832 . . . . . . . . . . . . . . . . . . . cubic millimeters per minute
43.8128 . . . . . . . . . . . . . . . . . . . . . . . . cubic millimeters per second

**GALLON PER HOUR: =**

0.0908506 . . . . . . . . . . . . . . . . . . . . . . . . . . . . kiloliters per day
0.00378544 . . . . . . . . . . . . . . . . . . . . . . . . . . . kiloliters per hour
0.0000630907 . . . . . . . . . . . . . . . . . . . . . . . . . kiloliters per minute
0.00000105151 . . . . . . . . . . . . . . . . . . . . . . . kiloliters per second
0.0908506 . . . . . . . . . . . . . . . . . . . . . . . . . . cubic meters per day
0.00378544 . . . . . . . . . . . . . . . . . . . . . . . . . cubic meters per hour
0.0000630907 . . . . . . . . . . . . . . . . . . . . . . cubic meters per minute
0.00000105151 . . . . . . . . . . . . . . . . . . . . . cubic meters per second
0.11882 . . . . . . . . . . . . . . . . . . . . . . . . . . . . cubic yards per day
0.004951 . . . . . . . . . . . . . . . . . . . . . . . . . . . cubic yards per hour
0.000082517 . . . . . . . . . . . . . . . . . . . . . . . . cubic yards per minute
0.0000013753 . . . . . . . . . . . . . . . . . . . . . . . cubic yards per second
0.57144 . . . . . . . . . . . . . . . . . . . . . . . . . . . . . . . barrels per day
0.0238095 . . . . . . . . . . . . . . . . . . . . . . . . . . . . . barrels per hour
0.00039683 . . . . . . . . . . . . . . . . . . . . . . . . . . . barrels per minute
0.0000066139 . . . . . . . . . . . . . . . . . . . . . . . . . barrels per second
0.908506 . . . . . . . . . . . . . . . . . . . . . . . . . . . . hektoliters per day
0.0378544 . . . . . . . . . . . . . . . . . . . . . . . . . . . hektoliters per hour
0.000630907 . . . . . . . . . . . . . . . . . . . . . . . . hektoliters per minute
0.0000105151 . . . . . . . . . . . . . . . . . . . . . . . hektoliters per second
2.57808 . . . . . . . . . . . . . . . . . . . . . . . bushels—U.S.—(dry) per day
0.10742 . . . . . . . . . . . . . . . . . . . . . . bushels—U.S.—(dry) per hour
0.0017903 . . . . . . . . . . . . . . . . . . . bushels—U.S.—(dry) per minute
0.000029839 . . . . . . . . . . . . . . . . . . bushels—U.S.—(dry) per second
2.49936 . . . . . . . . . . . . . . . . . . . . . . . bushels (Imperial) per day
0.10414 . . . . . . . . . . . . . . . . . . . . . . . bushels (Imperial) per hour
0.0017357 . . . . . . . . . . . . . . . . . . . . bushels (Imperial) per minute

GALLON PER HOUR (cont'd): =

| | |
|---|---|
| 0.000028928 | bushels (Imperial) per second |
| 3.20834 | cubic feet per day |
| 0.133681 | cubic feet per hour |
| 0.0022280 | cubic feet per minute |
| 0.000037134 | cubic feet per second |
| 9.0850560 | dekaliters per day |
| 0.378544 | dekaliters per hour |
| 0.00630907 | dekaliters per minute |
| 0.000105151 | dekaliters per second |
| 24 | gallons—U.S. (liquid) per day |
| 1.0 | gallons—U.S. (liquid) per hour |
| 0.016667 | gallons—U.S. (liquid) per minute |
| 0.00027778 | gallons—U.S. (liquid) per second |
| 19.98432 | gallons (Imperial) per day |
| 0.83268 | gallons (Imperial) per hour |
| 0.013878 | gallons (Imperial) per minute |
| 0.0002313 | gallons (Imperial) per second |
| 90.850560 | liters per day |
| 3.78544 | liters per hour |
| 0.0630907 | liters per minute |
| 0.00105151 | liters per second |
| 90.850560 | cubic decimeters per day |
| 3.785440 | cubic decimeters per hour |
| 0.0630907 | cubic decimeters per minute |
| 0.00105151 | cubic decimeters per second |
| 96 | quarts per day |
| 4.0 | quarts per hour |
| 0.066667 | quarts per minute |
| 0.0011111 | quarts per second |
| 192 | pints per day |
| 8.0 | pints per hour |
| 0.13333 | pints per minute |
| 0.0022222 | pints per second |
| 768 | gills per day |
| 32.0 | gills per hour |
| 0.53333 | gills per minute |
| 0.0088889 | gills per second |
| 908.505600 | deciliters per day |
| 37.854400 | deciliters per hour |
| 0.630907 | deciliters per minute |
| 0.0105151 | deciliters per second |
| 5,544 | cubic inches per day |
| 231 | cubic inches per hour |
| 3.85 | cubic inches per minute |

GALLON PER HOUR (cont'd): =

0.064167 . . . . . . . . . . . . . . . . . . . . . . . . . .cubic inches per second
9,085.055999 . . . . . . . . . . . . . . . . . . . . . centiliters per day
378.544000 . . . . . . . . . . . . . . . . . . . . . centiliters per hour
6.309067 . . . . . . . . . . . . . . . . . . . . . . centiliters per minute
0.105151 . . . . . . . . . . . . . . . . . . . . . . centiliters per second
90,850.559990 . . . . . . . . . . . . . . . . . . . milliliters per day
3,785.440 . . . . . . . . . . . . . . . . . . . . . .milliliters per hour
63.090667 . . . . . . . . . . . . . . . . . . . . . milliliters per minute
1.051511 . . . . . . . . . . . . . . . . . . . . . . milliliters per second
90,850.559990 . . . . . . . . . . . . . . . . .cubic centimeters per day
3,785.440 . . . . . . . . . . . . . . . . . . . cubic centimeters per hour
63.090667 . . . . . . . . . . . . . . . . . .cubic centimeters per minute
1.051511 . . . . . . . . . . . . . . . . . . .cubic centimeters per second
90,850,560 . . . . . . . . . . . . . . . . . . . cubic millimeters per day
3,785,440 . . . . . . . . . . . . . . . . . . . cubic millimeters per hour
63,090.666660 . . . . . . . . . . . . . . . .cubic millimeters per minute
1,051.511111 . . . . . . . . . . . . . . . . . cubic millimeters per second

GALLON PER MINUTE: =

5.451034 . . . . . . . . . . . . . . . . . . . . . . . . . . .kiloliters per day
0.227126 . . . . . . . . . . . . . . . . . . . . . . . . . . kiloliters per hour
0.00378544 . . . . . . . . . . . . . . . . . . . . . . . .kiloliters per minute
0.0000630907 . . . . . . . . . . . . . . . . . . . . . . kiloliters per second
5.451034 . . . . . . . . . . . . . . . . . . . . . . . cubic meters per day
0.227126 . . . . . . . . . . . . . . . . . . . . . . .cubic meters per hour
0.00378544 . . . . . . . . . . . . . . . . . . . . cubic meters per minute
0.0000630907 . . . . . . . . . . . . . . . . . . .cubic meters per second
7.129440 . . . . . . . . . . . . . . . . . . . . . . . . cubic yards per day
0.297060 . . . . . . . . . . . . . . . . . . . . . . . .cubic yards per hour
0.004951 . . . . . . . . . . . . . . . . . . . . . . cubic yards per minute
0.0000825167 . . . . . . . . . . . . . . . . . . . cubic yards per second
34.285680 . . . . . . . . . . . . . . . . . . . . . . . . . . . barrels per day
1.428570 . . . . . . . . . . . . . . . . . . . . . . . . . . barrels per hour
0.0238095 . . . . . . . . . . . . . . . . . . . . . . . . barrels per minute
0.000396825 . . . . . . . . . . . . . . . . . . . . . . barrels per second
54.510336 . . . . . . . . . . . . . . . . . . . . . . . .hektoliters per day
2.271264 . . . . . . . . . . . . . . . . . . . . . . . hektoliters per hour
0.0378544 . . . . . . . . . . . . . . . . . . . . .hektoliters per minute
0.000630907 . . . . . . . . . . . . . . . . . . . .hektoliters per second
154.6848 . . . . . . . . . . . . . . . . . . . .bushels—U.S. (dry) per day
6.4452 . . . . . . . . . . . . . . . . . . . . . bushels—U.S. (dry) per hour
0.10742 . . . . . . . . . . . . . . . . . . . .bushels—U.S. (dry) per minute
0.00179033 . . . . . . . . . . . . . . . . . . bushels—U.S. (dry) per second

GALLON PER MINUTE (cont'd): =

| | |
|---|---|
| 149.9616 | bushels (Imperial) per day |
| 6.2484 | bushels (Imperial) per hour |
| 0.10414 | bushels (Imperial) per minute |
| 0.00173567 | bushels (Imperial) per second |
| 192.500640 | cubic feet per day |
| 8.020860 | cubic feet per hour |
| 0.133681 | cubic feet per minute |
| 0.00222802 | cubic feet per second |
| 545.103359 | dekaliters per day |
| 22.712640 | dekaliters per hour |
| 0.378544 | dekaliters per minute |
| 0.00630907 | dekaliters per second |
| 1,440 | gallons—U.S. (liquid) per day |
| 60 | gallons—U.S. (liquid) per hour |
| 1.0 | gallons—U.S. (liquid) per minute |
| 0.0166667 | gallons—U.S. (liquid) per second |
| 1,199.05920 | gallons (Imperial) per day |
| 49.960800 | gallons (Imperial) per hour |
| 0.83268 | gallons (Imperial) per minute |
| 0.0138780 | gallons (Imperial) per second |
| 5,451.033594 | liters per day |
| 227.126400 | liters per hour |
| 3.78544 | liters per minute |
| 0.0630907 | liters per second |
| 5,451.033594 | cubic decimeters per day |
| 227.126400 | cubic decimeters per hour |
| 3.78544 | cubic decimeters per minute |
| 0.0630907 | cubic decimeters per second |
| 5,760 | quarts per day |
| 240 | quarts per hour |
| 4.0 | quarts per minute |
| 0.0666667 | quarts per second |
| 11,520 | pints per day |
| 480 | pints per hour |
| 8.0 | pints per minute |
| 0.133333 | pints per second |
| 46,080 | gills per day |
| 1,920 | gills per hour |
| 32.0 | gills per minute |
| 0.533333 | gills per second |
| 54,510.335942 | deciliters per day |
| 2,271.263998 | deciliters per hour |
| 37.854400 | deciliters per minute |
| 0.630907 | deciliters per second |

GALLON PER MINUTE (cont'd): =

| | |
|---|---|
| 332,640 | cubic inches per day |
| 13,860 | cubic inches per hour |
| 231.0 | cubic inches per minute |
| 3.850 | cubic inches per second |
| 545,103.359424 | centiliters per day |
| 22,712.639976 | centiliters per hour |
| 378.544000 | centiliters per minute |
| 6.309067 | centiliters per second |
| 5,451,034 | milliliters per day |
| 227,126.399760 | milliliters per hour |
| 3,785.440 | milliliters per minute |
| 63.090667 | milliliters per second |
| 5,451,034 | cubic centimeters per day |
| 227,126.399760 | cubic centimeters per hour |
| 3,785.440 | cubic centimeters per minute |
| 63.090667 | cubic centimeters per second |
| 5,451,033,594 | cubic millimeters per day |
| 227,126,400 | cubic millimeters per hour |
| 3,785,440 | cubic millimeters per minute |
| 63,090.6666 | cubic millimeters per second |

GALLON PER SECOND: =

| | |
|---|---|
| 327.062016 | kiloliters per day |
| 13.627584 | kiloliters per hour |
| 0.227126 | kiloliters per minute |
| 0.00378544 | kiloliters per second |
| 327.062016 | cubic meters per day |
| 13.627584 | cubic meters per hour |
| 0.227126 | cubic meters per minute |
| 0.00378544 | cubic meters per second |
| 427.766400 | cubic yards per day |
| 17.823600 | cubic yards per hour |
| 0.297060 | cubic yards per minute |
| 0.004951 | cubic yards per second |
| 2,057.140800 | barrels per day |
| 85.714200 | barrels per hour |
| 1.428570 | barrels per minute |
| 0.0238095 | barrels per second |
| 3,270.620160 | hektoliters per day |
| 136.275840 | hektoliters per hour |
| 2.271264 | hektoliters per minute |
| 0.0378544 | hektoliters per second |
| 9,281.08800 | bushels—U.S. (dry) per day |

GALLON PER SECOND (cont'd): =

| | |
|---|---|
| 386.712000 | bushels—U.S. (dry) per hour |
| 6.44520 | bushels—U.S. (dry) per minute |
| 0.10742 | bushels—U.S. (dry) per second |
| 8,997.696000 | bushels (Imperial) per day |
| 374.90400 | bushels (Imperial) per hour |
| 6.248400 | bushels (Imperial) per minute |
| 0.10414 | bushels (Imperial) per second |
| 11,550.038400 | cubic feet per day |
| 481.251600 | cubic feet per hour |
| 8.020860 | cubic feet per minute |
| 0.133681 | cubic feet per second |
| 32,706.201600 | dekaliters per day |
| 1,362.758400 | dekaliters per hour |
| 22.712640 | dekaliters per minute |
| 0.378544 | dekaliters per second |
| 86,400 | gallons—U.S. (liquid) per day |
| 3,600 | gallons—U.S. (liquid) per hour |
| 60 | gallons—U.S. (liquid) per minute |
| 1.0 | gallons—U.S. (liquid) per second |
| 71,943.55200 | gallons (Imperial) per day |
| 2.997.64800 | gallons (Imperial) per hour |
| 49.96080 | gallons (Imperial) per minute |
| 0.83268 | gallons (Imperial) per second |
| 327,062.01600 | liters per day |
| 13,627.58400 | liters per hour |
| 227.12640 | liters per minute |
| 3.78544 | liters per second |
| 327,062.01600 | cubic decimeters per day |
| 13,627.58400 | cubic decimeters per hour |
| 227.12640 | cubic decimeters per minute |
| 3.78544 | cubic decimeters per second |
| 345,600 | quarts per day |
| 14,400 | quarts per hour |
| 240 | quarts per minute |
| 4.0 | quarts per second |
| 691,200 | pints per day |
| 28,800 | pints per hour |
| 480 | pints per minute |
| 8.0 | pints per second |
| 2,764,800 | gills per day |
| 115,200 | gills per hour |
| 1,920 | gills per minute |
| 32.0 | gills per second |

GALLON PER SECOND (cont'd): =

| | |
|---|---|
| 3,270,620 | deciliters per day |
| 136,275.8400 | deciliters per hour |
| 2,271.2640 | deciliters per minute |
| 37.8544 | deciliters per second |
| 19,958,400 | cubic inches per day |
| 831,600 | cubic inches per hour |
| 13,860 | cubic inches per minute |
| 231 | cubic inches per second |
| 32,706,202 | centiliters per day |
| 1,362,758 | centiliters per hour |
| 22,712.640 | centiliters per minute |
| 378.544 | centiliters per second |
| 327,062,016 | milliliters per day |
| 13,627,584 | milliliters per hour |
| 227,126.40 | milliliters per minute |
| 3,785.44 | milliliters per second |
| 327,062,016 | cubic centimeters per day |
| 13,627,584 | cubic centimeters per hour |
| 227,126.40 | cubic centimeters per minute |
| 3,785.44 | cubic centimeters per second |
| 327,061,016,000 | cubic millimeters per day |
| 13,627,584,000 | cubic millimeters per hour |
| 227,126,400 | cubic millimeters per minute |
| 3,785,440 | cubic millimeters per second |

GRAIN (AVOIRDUPOIS): =

| | |
|---|---|
| 0.000000637755089 | tons (long) |
| 0.000000647989857 | tons (metric) |
| 0.0000007142857 | tons (short) |
| 0.000064798918 | kilograms |
| 0.00014285714 | pounds (Avoir.) |
| 0.00017361111 | pounds (Troy) |
| 0.00064798918 | hektograms |
| 0.0064798918 | dekagrams |
| 0.00208333 | ounces (Troy) |
| 0.0022857 | ounces (Avoir.) |
| 0.03657143 | drams (Avoir.) |
| 0.0166667 | drams (Troy) |
| 0.0416667 | pennyweights (Troy) |
| 0.05000 | scruples (Troy) |
| 0.064798918 | grams |
| 0.64798918 | decigrams |

GRAIN (AVOIRDUPOIS) (cont'd): =

| | |
|---|---|
| 6.4798918 | centigrams |
| 64.798918 | milligrams |
| 0.3240 | carats (metric) |

GRAIN (AVOIR.) PER BARREL: =

| | |
|---|---|
| 718.956 | parts per million |
| 0.000793548 | tons (net) per cubic meter |
| 0.71988 | kilograms per cubic meter |
| 1.587054 | pounds (Avoir.) per cubic meter |
| 25.39320 | ounces (Avoir.) per cubic meter |
| 406.29120 | drams (Avoir.) per cubic meter |
| 719.88 | grams per cubic meter |
| 0.000126161 | tons (net) per barrel |
| 0.114449 | kilograms per barrel |
| 0.252316 | pounds (Avoir.) per barrel |
| 4.037124 | ounces (Avoir.) per barrel |
| 64.59390 | drams (Avoir.) per barrel |
| 114.44916 | grams per barrel |
| 0.00000300384 | tons (net) per gallon—U.S. |
| 0.00272498 | kilograms per gallon—U.S. |
| 0.00600753 | pounds (Avoir.) per gallon—U.S. |
| 0.096122 | ounces (Avoir.) per gallon—U.S. |
| 1.53795 | drams (Avoir.) per gallon—U.S. |
| 2.724982 | grams per gallon—U.S. |
| 0.000000793548 | tons (net) per liter |
| 0.00071988 | kilograms per liter |
| 0.00158705 | pounds (Avoir.) per liter |
| 0.0253932 | ounces (Avoir.) per liter |
| 0.406291 | drams (Avoir.) per liter |
| 0.71988 | grams per liter |
| 0.000000000793548 | tons (net) per cubic centimeter |
| 0.00000071980 | kilograms per cubic centimeter |
| 0.00000158705 | pounds (Avoir.) per cubic centimeter |
| 0.0000253932 | ounces (Avoir.) per cubic centimeter |
| 0.000406291 | drams (Avoir.) per cubic centimeter |
| 0.00071988 | grams per cubic centimeter |

GRAIN (AVOIR.) PER CUBIC CENTIMETER: =

| | |
|---|---|
| 0.00452219 | parts per million |
| 0.00000000499137 | tons (net) per cubic meter |

## GRAIN (AVOIR.) PER CUBIC CENTIMETER (cont'd): =

| | |
|---|---|
| 0.00000452800 | kilograms per cubic meter |
| 0.00000998248 | pounds (Avoir.) per cubic meter |
| 0.000159722 | ounces (Avoir.) per cubic meter |
| 0.00255555 | drams (Avoir.) per cubic meter |
| 0.00452800 | grams |
| 0.000000000793545 | tons (net) per barrel |
| 0.000000719878 | kilograms per barrel |
| 0.00000158705 | pounds (Avoir.) per barrel |
| 0.0000253933 | ounces (Avoir.) per barrel |
| 0.000406292 | drams (Avoir.) per barrel |
| 0.000719879 | grams per barrel |
| 0.0000000000188940 | tons (net) per gallon—U.S. |
| 0.00000001714 | kilograms per gallon—U.S. |
| 0.0000000377873 | pounds (Avoir.) per gallon—U.S. |
| 0.000000604602 | ounces (Avoir.) per gallon—U.S. |
| 0.00000967362 | drams (Avoir.) per gallon—U.S. |
| 0.00001714 | grams per gallon—U.S. |
| 0.00000000000499137 | tons (net) per liter |
| 0.00000000452800 | kilograms per liter |
| 0.00000000998248 | pounds (Avoir.) per liter |
| 0.000000159722 | ounces (Avoir.) per liter |
| 0.00000255555 | drams (Avoir.) per liter |
| 0.00000452800 | grams per liter |
| 0.00000000000000499137 | tons (net) per cubic centimeter |
| 0.00000000000452800 | kilograms per cubic centimeter |
| 0.00000000000998248 | pounds (Avoir.) per cubic centimeter |
| 0.000000000159722 | ounces (Avoir.) per cubic centimeter |
| 0.00000000255555 | drams (Avoir.) per cubic centimeter |
| 0.00000000452800 | grams per cubic centimeter |

## GRAIN (AVOIR.) PER CUBIC METER: =

| | |
|---|---|
| 4,522.192260 | parts per million |
| 0.00499137 | tons (net) per cubic meter |
| 4.528004 | kilograms per cubic meter |
| 9.982479 | pounds (Avoir.) per cubic meter |
| 159.721781 | ounces (Avoir.) per cubic meter |
| 2,555.548489 | drams (Avoir.) per cubic meter |
| 4,528.004167 | grams per cubic meter |
| 0.000793545 | tons (net) per barrel |
| 0.719878 | kilograms per barrel |
| 1.587053 | pounds (Avoir.) per barrel |
| 25.393280 | ounces (Avoir.) per barrel |

GRAIN (AVOIR.) PER CUBIC METER: (cont'd): =

| | |
|---|---|
| 406.291949 | drams (Avoir.) per barrel |
| 719.878693 | grams per barrel |
| 0.0000188940 | tons (net) per gallon—U.S. |
| 0.01714 | kilograms per gallon—U.S. |
| 0.0377873 | pounds (Avoir.) per gallon—U.S. |
| 0.604602 | ounces (Avoir.) per gallon—U.S. |
| 9.673618 | drams (Avoir.) per gallon—U.S. |
| 17.14 | grams per gallon—U.S. |
| 0.00000499137 | tons (net) per liter |
| 0.00452800 | kilograms per liter |
| 0.00998248 | pounds (Avoir.) per liter |
| 0.159722 | ounces (Avoir.) per liter |
| 2.555548 | drams (Avoir.) per liter |
| 4.528004 | grams per liter |
| 0.00000000499137 | tons (net) per cubic centimeter |
| 0.00000452800 | kilograms per cubic centimeter |
| 0.00000998248 | pounds (Avoir.) per cubic centimeter |
| 0.000159722 | ounces (Avoir.) per cubic centimeter |
| 0.00255555 | drams (Avoir.) per cubic centimeter |
| 0.00452800 | grams |

GRAIN (AVOIR.) PER GALLON—U.S.: =

| | |
|---|---|
| 17.118 | parts per million |
| 0.000018894 | tons (net) per cubic meter |
| 0.01714 | kilograms per cubic meter |
| 0.037787 | pounds (Avoir.) per cubic meter |
| 0.60460 | ounces (Avoir.) per cubic meter |
| 9.67360 | drams (Avoir.) per cubic meter |
| 17.14 | grams per cubic meter |
| 0.00000300384 | tons (net) per barrel |
| 0.00272498 | kilograms per barrel |
| 0.00600753 | pounds (Avoir.) per barrel |
| 0.096122 | ounces (Avoir.) per barrel |
| 1.53795 | drams (Avoir.) per barrel |
| 2.72498 | grams per barrel |
| 0.000000071520 | tons (net) per gallon—U.S. |
| 0.000064881 | kilograms per gallon—U.S. |
| 0.00014304 | pounds (Avoir.) per gallon—U.S. |
| 0.0022886 | ounces (Avoir.) per gallon—U.S. |
| 0.036617 | drams (Avoir.) per gallon—U.S. |
| 0.064881 | grams per gallon—U.S. |
| 0.000000018894 | tons (net) per liter |

## GRAIN (AVOIR) PER GALLON–U.S. (cont'd): =

| | |
|---|---|
| 0.00001714 | kilograms per liter |
| 0.000037787 | pounds (Avoir.) per liter |
| 0.00060460 | ounces (Avoir.) per liter |
| 0.0096735 | drams (Avoir.) per liter |
| 0.01714 | grams per liter |
| 0.000000000018894 | tons (net) per cubic centimeter |
| 0.00000001714 | kilograms per cubic centimeter |
| 0.000000037787 | pounds (Avoir.) per cubic centimeter |
| 0.00000060460 | ounces (Avoir.) per cubic centimeter |
| 0.0000096735 | drams (Avoir.) per cubic centimeter |
| 0.00001714 | grams per cubic centimeter |

## GRAIN (AVOIR.) PER LITER: =

| | |
|---|---|
| 4.522192 | parts per million |
| 0.00000499137 | tons (net) per cubic meter |
| 0.00452800 | kilograms per cubic meter |
| 0.00998248 | pounds (Avoir.) per cubic meter |
| 0.159722 | ounces (Avoir.) per cubic meter |
| 2.555548 | drams (Avoir.) per cubic meter |
| 4.52804 | grams per cubic meter |
| 0.00000793545 | tons (net) per barrel |
| 0.000719878 | kilograms per barrel |
| 0.00158705 | pounds (Avoir.) per barrel |
| 0.0253933 | ounces (Avoir.) per barrel |
| 0.406292 | drams (Avoir.) per barrel |
| 0.719879 | grams per barrel |
| 0.0000000188940 | tons (net) per gallon–U.S. |
| 0.00001714 | kilograms per gallon–U.S. |
| 0.0000377873 | pounds (Avoir.) per gallon–U.S. |
| 0.00604602 | ounces (Avoir.) per gallon–U.S. |
| 0.00967362 | drams (Avoir.) per gallon–U.S. |
| 0.01714 | grams per gallon–U.S. |
| 0.00000000499137 | tons (net) per liter |
| 0.00000452800 | kilograms per liter |
| 0.00000998248 | pounds (Avoir.) per liter |
| 0.000159722 | ounces (Avoir.) per liter |
| 0.00255555 | drams (Avoir.) per liter |
| 0.00452800 | grams per liter |
| 0.00000000000499137 | tons (net) per cubic centimeter |
| 0.00000000452800 | kilograms per cubic centimeter |
| 0.00000000998248 | pounds (Avoir.) per cubic centimeter |
| 0.000000159722 | ounces (Avoir.) per cubic centimeter |

GRAIN (AVOIR.) PER LITER (cont'd): =

| | |
|---|---|
| 0.00000255555 | drams (Avoir.) per cubic centimeter |
| 0.00000452800 | grams per cubic centimeter |

GRAM: =

| | |
|---|---|
| 0.00000098426 | tons (long) |
| 0.000001 | tons (metric) |
| 0.00000110231 | tons (short) |
| 0.001 | kilograms |
| 0.00220462 | pounds (Avoir.) |
| 0.00267923 | pounds (Troy) |
| 0.01 | hektograms |
| 0.1 | dekagrams |
| 0.0321507 | ounces (Troy) |
| 0.03527392 | ounces (Avoir.) |
| 0.257206 | drams (Troy) |
| 0.564383 | drams (Avoir.) |
| 0.6430149 | pennyweights |
| 0.771618 | scruples |
| 5.0 | carats (metric) |
| 10.0 | decigrams |
| 15.4324 | grains |
| 100 | centigrams |
| 1,000 | milligrams |

GRAM CALORIE (MEAN): =

| | |
|---|---|
| 99.334 | foot poundals |
| 1,192.008 | inch poundals |
| 3.0874 | foot pounds |
| 37.0488 | inch pounds |
| 0.0000001 | ton (short) calories |
| 0.001 | kilogram calories |
| 0.00204622 | pound calories |
| 0.01 | hektogram calories |
| 0.1 | dekagram calories |
| 0.0327395 | ounce calories |
| 1.0 | gram calorie |
| 10.0 | decigram calories |
| 100.0 | centigram calories |
| 1,000 | milligram calories |
| 0.00000004845 | kilowatt days |

GRAM CALORIE (MEAN) (cont'd): =

| | |
|---|---|
| 0.0000011628 | kilowatt hours |
| 0.0000697680 | kilowatt minutes |
| 0.004186 | kilowatt seconds |
| 0.00004845 | watt days |
| 0.0011628 | watt hours |
| 0.0697680 | watt minutes |
| 4.186 | watt seconds |
| 0.00042685 | ton meters |
| 0.42685 | kilogram meters |
| 0.941043 | pound meters |
| 4.2685 | hektogram meters |
| 42.685 | dekagram meters |
| 15.056688 | ounce meters |
| 426.85 | gram meters |
| 4,268.5 | decigram meters |
| 42,685 | centigram meters |
| 426,850 | milligram meters |
| 0.0000042685 | ton hektometers |
| 0.0042685 | kilogram hektometers |
| 0.00941043 | pound hektometers |
| 0.042685 | hektogram hektometers |
| 0.42685 | dekagram hektometers |
| 0.150567 | ounce hektometers |
| 4.2685 | gram hektometers |
| 42.685 | decigram hektometers |
| 426.85 | centigram hektometers |
| 4,268.5 | milligram hektometers |
| 0.000042685 | ton dekameters |
| 0.042685 | kilogram dekameters |
| 0.0941043 | pound dekameters |
| 0.42685 | hektogram dekameters |
| 4.2685 | dekagram dekameters |
| 1.50567 | ounce dekameters |
| 42.685 | gram dekameters |
| 426.85 | decigram dekameters |
| 4,268.5 | centigram dekameters |
| 42,685 | milligram dekameters |
| 0.00140042 | ton feet |
| 1.400424 | kilogram feet |
| 0.0308704 | pound feet |
| 14.004236 | hektogram feet |
| 140.042357 | dekagram feet |
| 0.493985 | ounce feet |
| 1,400.423566 | gram feet |

GRAM CALORIE (MEAN) (cont'd): =

| | |
|---|---|
| 14,004.235661 | decigram feet |
| 140,042.356605 | centigram feet |
| 1,400,423.0 | milligram feet |
| 0.0168050 | ton inches |
| 16.805083 | kilogram inches |
| 37.0488557 | pound inches |
| 168.050828 | hektogram inches |
| 1,680.0508284 | dekagram inches |
| 592.78169 | ounce inches |
| 16,805.082792 | gram inches |
| 168,050.82792 | decigram inches |
| 1,680,508 | centigram inches |
| 16,805,082 | milligram inches |
| 0.0042685 | ton decimeters |
| 4.2685 | kilogram decimeters |
| 9.410430 | pound decimeters |
| 42.685 | hektogram decimeters |
| 426.85 | dekagram decimeters |
| 150.566885 | ounce decimeters |
| 4,268.5 | gram decimeters |
| 42,685 | decigram decimeters |
| 426,850 | centigram decimeters |
| 4,268,500 | milligram decimeters |
| 0.042685 | ton centimeters |
| 42.685 | kilogram centimeters |
| 94.104303 | pound centimeters |
| 426.85 | hektogram centimeters |
| 4,268.5 | dekagram centimeters |
| 1,505.668846 | ounce centimeters |
| 42,685.0 | gram centimeters |
| 426,850 | decigram centimeters |
| 4,268,500 | centigram centimeters |
| 42,685,000 | milligram centimeters |
| 0.42685 | ton millimeters |
| 426.85 | kilogram millimeters |
| 941.043029 | pound millimeters |
| 4,268.5 | hektogram millimeters |
| 42,685 | dekagram millimeters |
| 15,056.688464 | ounce millimeters |
| 426,850 | gram millimeters |
| 4,268,500 | decigram millimeters |
| 42,685,000 | centigram millimeters |
| 426,850,000 | milligram millimeters |
| 0.000041311 | kiloliter-atmospheres |

GRAM CALORIE (MEAN) (cont.d): =

0.00041311 . . . . . . . . . . . . . . . . . . . . . . . . hektoliter-atmospheres
0.0041311 . . . . . . . . . . . . . . . . . . . . . . . . . . dekaliter-atmospheres
0.041311 . . . . . . . . . . . . . . . . . . . . . . . . . . . .liter-atmospheres
0.41311 . . . . . . . . . . . . . . . . . . . . . . . . . . . deciliter-atmospheres
4.1311 . . . . . . . . . . . . . . . . . . . . . . . . . . . .centiliter-atmospheres
41.311 . . . . . . . . . . . . . . . . . . . . . . . . . . . . milliliter-atmospheres
0.0000000000000415977 . . . . . . . . . . . . cubic kilometer-atmospheres
0.0000000000415977 . . . . . . . . . . . . . .cubic hektometer-atmospheres
0.0000000415977 . . . . . . . . . . . . . . . . . cubic dekameter-atmospheres
0.0000415977 . . . . . . . . . . . . . . . . . . . . . cubic meter-atmospheres
0.001469 . . . . . . . . . . . . . . . . . . . . . . . . . cubic feet-atmospheres
2.538415 . . . . . . . . . . . . . . . . . . . . . . . . . cubic inch-atmospheres
0.0415977 . . . . . . . . . . . . . . . . . . . . . .cubic decimeter-atmospheres
41.597673 . . . . . . . . . . . . . . . . . . . . . cubic centimeter-atmospheres
41,597.673 . . . . . . . . . . . . . . . . . . . . . cubic millimeter-atmospheres
4.185829 . . . . . . . . . . . . . . . . . . . . . . . . . . . . . . .joules per gram
0.00000158097 . . . . . . . . . . . . . . . . . . . . . . . Cheval-Vapeur hours
0.0000000649708 . . . . . . . . . . . . . . . . . . . . . . . . .horsepower days
0.0000015593 . . . . . . . . . . . . . . . . . . . . . . . . . . horsepower hours
0.0000935980 . . . . . . . . . . . . . . . . . . . . . . . . .horsepower minutes
0.00561588 . . . . . . . . . . . . . . . . . . . . . . . . . . horsepower seconds
0.000000271842 . . . . . . . . . . . . . . . . . . pounds of carbon oxidized with 100% efficiency
0.00000408756 . . . . . . . . . . . . . . . . pounds of water evaporated from and at 212° F.
1.8 . . . . . . . . . . . . . . . . . . . . . . . . . . . . . . . BTU (mean) per pound
0.0039685 . . . . . . . . . . . . . . . . . . . . . . . . . . . . . . . . . . . . . .BTU
41,858,291 . . . . . . . . . . . . . . . . . . . . . . . . . . . . . . . . . . . . . ergs

GRAM CALORIE:
(15° C per square centimeter per second for temperature gradient of 1° C per centimeter): =

4.185829 . . . . . . . . . . . . . . . . . . . . . . . .joules (absolute) per square centimeter per second for temperature gradient of 1° C per centimeter
0.80620 . . . . . . . . . . . . . . . . . . . . . . . . . .BTU (mean) per square foot per second for a temperature gradient of 1° F. per inch

## GRAM WEIGHT SECOND PER SQUARE CENTIMETER: =

| | |
|---|---|
| 980.665 | poises |
| 98,066.5 | centipoises |

## GRAM PER CUBIC CENTIMETER: =

| | |
|---|---|
| 100,000 | kilograms per cubic meter |
| 2,831.7 | kilograms per cubic foot |
| 16.38776 | kilograms per cubic inch |
| 0.001 | kilograms per cubic centimeter |
| 2,204.62 | pounds per cubic meter |
| 62.42822 | pounds per cubic foot |
| 0.036127 | pounds per cubic inch |
| 0.0022046 | pounds per cubic centimeter |
| 35,273.92 | ounces per cubic meter |
| 998.85159 | ounces per cubic foot |
| 0.57804 | ounces per cubic inch |
| 0.035274 | ounces per cubic centimeter |
| 1,000,000 | grams per cubic meter |
| 28,317.0 | grams per cubic foot |
| 16.38705 | grams per cubic inch |
| 1.0 | grams per cubic centimeter |
| 15,432,400 | grains per cubic meter |
| 436,999.27080 | grains per cubic foot |
| 252.89148 | grains per cubic inch |
| 15.4324 | grains per cubic centimeter |

## GRAM PER CUBIC FOOT: =

| | |
|---|---|
| 0.035314 | kilograms per cubic meter |
| 0.001 | kilograms per cubic foot |
| 0.00000057870 | kilograms per cubic inch |
| 0.000000035314 | kilograms per cubic centimeter |
| 0.077854 | pounds per cubic meter |
| 0.0022046 | pounds per cubic foot |
| 0.0000012758 | pounds per cubic inch |
| 0.000000077854 | pounds per cubic centimeter |
| 1.24568 | ounces per cubic meter |
| 0.035274 | ounces per cubic foot |
| 0.000020413 | ounces per cubic inch |
| 0.00000124578 | ounces per cubic centimeter |
| 35.31444 | grams per cubic meter |
| 1.0 | grams per cubic foot |

## GRAM PER CUBIC FOOT: (cont'd): =

0.00057870 . . . . . . . . . . . . . . . . . . . . . . . . . . . .grams per cubic inch
0.0000353144 . . . . . . . . . . . . . . . . . . . . . . . . grams per cubic centimeter
544.98657 . . . . . . . . . . . . . . . . . . . . . . . . . . . .grains per cubic meter
15.4324 . . . . . . . . . . . . . . . . . . . . . . . . . . . . . . .grains per cubic foot
0.0089307 . . . . . . . . . . . . . . . . . . . . . . . . . . . . .grains per cubic inch
0.00054499 . . . . . . . . . . . . . . . . . . . . . . . . . grains per cubic centimeter

## GRAM PER CUBIC INCH: =

61.203 . . . . . . . . . . . . . . . . . . . . . . . . . . kilograms per cubic meter
1.72799 . . . . . . . . . . . . . . . . . . . . . . . . . . . kilograms per cubic foot
0.001 . . . . . . . . . . . . . . . . . . . . . . . . . . . . kilograms per cubic inch
0.000061203 . . . . . . . . . . . . . . . . . . . . . .kilograms per cubic centimeter
134.53131 . . . . . . . . . . . . . . . . . . . . . . . . . . . pounds per cubic meter
3.80952 . . . . . . . . . . . . . . . . . . . . . . . . . . . . . .pounds per cubic foot
0.0022046 . . . . . . . . . . . . . . . . . . . . . . . . . . . .pounds per cubic inch
0.00013453 . . . . . . . . . . . . . . . . . . . . . . . pounds per cubic centimeter
2,152.52530 . . . . . . . . . . . . . . . . . . . . . . . . . ounces per cubic meter
60.95306 . . . . . . . . . . . . . . . . . . . . . . . . . . . . ounces per cubic foot
0.035274 . . . . . . . . . . . . . . . . . . . . . . . . . . . . ounces per cubic inch
0.0021525 . . . . . . . . . . . . . . . . . . . . . . . . ounces per cubic centimeter
610,230 . . . . . . . . . . . . . . . . . . . . . . . . . . . . . .grams per cubic meter
17,279.88291 . . . . . . . . . . . . . . . . . . . . . . . . . . .grams per cubic foot
1.0 . . . . . . . . . . . . . . . . . . . . . . . . . . . . . . . . . . .grams per cubic inch
0.61023 . . . . . . . . . . . . . . . . . . . . . . . . . . grams per cubic centimeter
941,731.3452 . . . . . . . . . . . . . . . . . . . . . . . . . .grains per cubic meter
26,667.0065002 . . . . . . . . . . . . . . . . . . . . . . . .grains per cubic foot
15.4324 . . . . . . . . . . . . . . . . . . . . . . . . . . . . . . .grains per cubic inch
0.94173 . . . . . . . . . . . . . . . . . . . . . . . . . . grains per cubic centimeter

## GRAM PER CUBIC METER: =

0.001 . . . . . . . . . . . . . . . . . . . . . . . . . . . kilograms per cubic meter
0.00028317 . . . . . . . . . . . . . . . . . . . . . . . . . kilograms per cubic foot
0.00000016387 . . . . . . . . . . . . . . . . . . . . . . kilograms per cubic inch
0.000000001 . . . . . . . . . . . . . . . . . . . . . .kilograms per cubic centimeter
0.00220462 . . . . . . . . . . . . . . . . . . . . . . . . . . pounds per cubic meter
0.000062428 . . . . . . . . . . . . . . . . . . . . . . . . . . .pounds per cubic foot
0.000000036127 . . . . . . . . . . . . . . . . . . . . . . . .pounds per cubic inch
0.00000000220462 . . . . . . . . . . . . . . . . . . pounds per cubic centimeter
0.035274 . . . . . . . . . . . . . . . . . . . . . . . . . . . . ounces per cubic meter
0.00099885 . . . . . . . . . . . . . . . . . . . . . . . . . . . ounces per cubic foot

## GRAM PER CUBIC METER (cont'd): =

| | |
|---|---|
| 0.00000057803 | ounces per cubic inch |
| 0.000000035274 | ounces per cubic centimeter |
| 1.0 | grams per cubic meter |
| 0.028317 | grams per cubic foot |
| 0.000016387 | grams per cubic inch |
| 0.000001 | grams per cubic centimeter |
| 15.4324 | grains per cubic meter |
| 0.437 | grains per cubic foot |
| 0.00025289 | grains per cubic inch |
| 0.000015434 | grains per cubic centimeter |

## GRAM PER CENTIMETER: =

| | |
|---|---|
| 0.100 | kilograms per meter |
| 0.03048006 | kilograms per foot |
| 0.00254005 | kilograms per inch |
| 0.001 | kilograms per centimeter |
| 0.22046 | pounds per meter |
| 0.067197 | pounds per foot |
| 0.0054014 | pounds per inch |
| 0.0022046 | pounds per centimeter |
| 3.52736 | ounces per meter |
| 1.075152 | ounces per foot |
| 0.086422 | ounces per inch |
| 0.035274 | ounces per centimeter |
| 100 | grams per meter |
| 30.48006 | grams per foot |
| 2.54005 | grams per inch |
| 1.0 | grams per centimeter |
| 1,543.24 | grains per meter |
| 470.38048 | grains per foot |
| 39.19822 | grains per inch |
| 15.4324 | grains per centimeter |

## GRAM PER FOOT: =

| | |
|---|---|
| 0.0032808 | kilograms per meter |
| 0.001 | kilograms per foot |
| 0.000083333 | kilograms per inch |
| 0.000032808 | kilograms per centimeter |
| 0.0072329 | pounds per meter |
| 0.0022046 | pounds per foot |

GRAM PER FOOT (cont'd): =

0.00018372 . . . . . . . . pounds per inch
0.000072329 . . . . . . . . pounds per centimeter
0.11573 . . . . . . . . ounces per meter
0.035274 . . . . . . . . ounces per foot
0.0029395 . . . . . . . . ounces per inch
0.0011573 . . . . . . . . ounces per centimeter
3.28083 . . . . . . . . grams per meter
1.0 . . . . . . . . grams per foot
0.083333 . . . . . . . . grams per inch
0.0328083 . . . . . . . . grams per centimeter
50.63113 . . . . . . . . grains per meter
15.4324 . . . . . . . . grains per foot
1.28603 . . . . . . . . grains per inch
0.50631 . . . . . . . . grains per centimeter

GRAM PER INCH: =

0.03937 . . . . . . . . kilograms per meter
0.012 . . . . . . . . kilograms per foot
0.001 . . . . . . . . kilograms per inch
0.00039370 . . . . . . . . kilograms per centimeter
0.086796 . . . . . . . . pounds per meter
0.026455 . . . . . . . . pounds per foot
0.0022046 . . . . . . . . pounds per inch
0.00086796 . . . . . . . . pounds per centimeter
1.38874 . . . . . . . . ounces per meter
0.42329 . . . . . . . . ounces per foot
0.035274 . . . . . . . . ounces per inch
0.013887 . . . . . . . . ounces per centimeter
39.37000 . . . . . . . . grams per meter
12.0 . . . . . . . . grams per foot
1.0 . . . . . . . . grams per inch
0.39370 . . . . . . . . grams per centimeter
607.57359 . . . . . . . . grains per meter
185.1888 . . . . . . . . grains per foot
15.4324 . . . . . . . . grains per inch
6.075736 . . . . . . . . grains per centimeter

GRAM PER LITER: =

8.34543 . . . . . . . . pounds per 1000 gallons
1,000 . . . . . . . . parts per million

GRAM PER LITER (cont'd); =

|  |  |
|---|---|
| 0.00110230 | tons (net) per cubic meter |
| 1.0 | kilograms per cubic meter |
| 2.2046099 | pounds (Avoir.) per cubic meter |
| 35.273758 | ounces (Avoir.) per cubic meter |
| 564.379806 | drams (Avoir.) per cubic meter |
| 1,000 | grams per cubic meter |
| 15,432.258039 | grains per cubic meter |
| 0.0000312140 | tons (net) per cubic foot |
| 0.0283169 | kilograms per cubic foot |
| 0.0624280 | pounds (Avoir.) per cubic foot |
| 0.998848 | ounces (Avoir.) per cubic foot |
| 15.981559 | drams (Avoir.) per cubic foot |
| 28.316846 | grams per cubic foot |
| 436.995689 | grains per cubic foot |
| 0.000175254 | tons (net) per barrel |
| 0.158988 | kilograms per barrel |
| 0.350508 | pounds (Avoir.) per barrel |
| 5.608134 | ounces (Avoir.) per barrel |
| 89.730060 | drams (Avoir.) per barrel |
| 158.987766 | grams per barrel |
| 2,543.556 | grains per barrel |
| 0.00000417271 | tons (net) per gallon—U.S. |
| 0.00378542 | kilograms per gallon—U.S. |
| 0.00834543 | pounds (Avoir.) per gallon—U.S. |
| 0.133527 | ounces (Avoir.) per gallon—U.S. |
| 2.136430 | drams (Avoir.) per gallon—U.S. |
| 3.785423 | grams per gallon—U.S. |
| 58.418 | grains per gallon—U.S. |
| 0.00000110231 | tons (net) per liter |
| 0.001 | kilograms per liter |
| 0.00220462 | pounds (Avoir.) per liter |
| 0.0352739 | ounces (Avoir.) per liter |
| 0.564383 | drams (Avoir.) per liter |
| 15.4324 | grains per liter |
| 0.00000000110231 | tons (net) per cubic centimeter |
| 0.000001 | kilograms per cubic centimeter |
| 0.00000220462 | pounds (Avoir.) per cubic centimeter |
| 0.0000352739 | ounces (Avoir.) per cubic centimeter |
| 0.000564383 | drams (Avoir.) per cubic centimeter |
| 0.001 | grams per cubic centimeter |
| 0.0154324 | grains per cubic centimeter |

## GRAM PER METER: =

| | |
|---|---|
| 0.001 | kilograms per meter |
| 0.00030480 | kilograms per foot |
| 0.000025401 | kilograms per inch |
| 0.00001 | kilograms per centimeter |
| 0.0022046 | pounds per meter |
| 0.00067197 | pounds per foot |
| 0.0000560 | pounds per inch |
| 0.000022046 | pounds per centimeter |
| 0.035274 | ounces per meter |
| 0.010752 | ounces per foot |
| 0.0008960 | ounces per inch |
| 0.00035274 | ounces per centimeter |
| 1.0 | grams per meter |
| 0.30480 | grams per foot |
| 0.025401 | grams per inch |
| 0.01 | grams per centimeter |
| 15.4324 | grains per meter |
| 4.70380 | grains per foot |
| 0.39198 | grains per inch |
| 0.15432 | grains per centimeter |

## GRAM PER SQUARE CENTIMETER: =

| | |
|---|---|
| 10 | kilograms per square meter |
| 0.92903 | kilograms per square foot |
| 0.0064516 | kilograms per square inch |
| 0.001 | kilograms per square centimeter |
| 22.046 | pounds per square meter |
| 2.048140 | pounds per square foot |
| 0.014223 | pounds per square inch |
| 0.0022046 | pounds per square centimeter |
| 352.7392 | ounces per square meter |
| 32.77053 | ounces per square foot |
| 0.22756 | ounces per square inch |
| 0.035274 | ounces per square centimeter |
| 10,000 | grams per square meter |
| 929.03 | grams per square foot |
| 6.45156 | grams per square inch |
| 1.0 | grams per square centimeter |
| 154,324 | grains per square meter |
| 14,337.16257 | grains per square foot |
| 99.56299 | grains per square inch |
| 15.4324 | grains per square centimeter |

GRAM PER SQUARE CENTIMETER: (cont'd): =

| | |
|---|---|
| 0.73556 | millimeters of mercury @ 0° C |
| 0.00096784 | atmospheres |
| 980.665 | dynes |

GRAM PER SQUARE FOOT: =

| | |
|---|---|
| 0.010764 | kilograms per square meter |
| 0.001 | kilograms per square foot |
| 0.0000069444 | kilograms per square inch |
| 0.0000010764 | kilograms per square centimeter |
| 0.023730 | pounds per square meter |
| 0.00220462 | pounds per square foot |
| 0.000015310 | pounds per square inch |
| 0.0000023730 | pounds per square centimeter |
| 0.37968 | ounces per square meter |
| 0.035274 | ounces per square foot |
| 0.00024496 | ounces per square inch |
| 0.000037968 | ounces per square centimeter |
| 10.76387 | grams per square meter |
| 1.0 | grams per square foot |
| 0.0069444 | grams per square inch |
| 0.00107639 | grams per square centimeter |
| 166.11235 | grains per square meter |
| 15.4324 | grains per square foot |
| 0.10717 | grains per square inch |
| 0.016611 | grains per square centimeter |

GRAM PER SQUARE INCH: =

| | |
|---|---|
| 1.55 | kilograms per square meter |
| 0.1440 | kilograms per square foot |
| 0.001 | kilograms per square inch |
| 0.000155 | kilograms per square centimeter |
| 3.41713 | pounds per square meter |
| 0.31746 | pounds per square foot |
| 0.0022046 | pounds per square inch |
| 0.00034171 | pounds per square centimeter |
| 54.67470 | ounces per square meter |
| 5.0794444 | ounces per square foot |
| 0.035274 | ounces per square inch |
| 0.0054675 | ounces per square centimeter |
| 1,550 | grams per square meter |
| 143.99965 | grams per square foot |

## GRAM PER SQUARE INCH (cont'd): =

| | |
|---|---|
| 1.0 | grams per square inch |
| 0.155 | grams per square centimeter |
| 23,920.099878 | grains per square meter |
| 2,222.25834 | grains per square foot |
| 15.4324 | grains per square inch |
| 2.39201 | grains per square centimeter |

## GRAM PER SQUARE METER: =

| | |
|---|---|
| 0.001 | kilograms per square meter |
| 0.000092903 | kilograms per square foot |
| 0.00000064516 | kilograms per square inch |
| 0.0000001 | kilograms per square centimeter |
| 0.0022046 | pounds per square meter |
| 0.00020481 | pounds per square foot |
| 0.0000014222 | pounds per square inch |
| 0.00000022046 | pounds per square centimeter |
| 0.035274 | ounces per square meter |
| 0.0032771 | ounces per square foot |
| 0.000022757 | ounces per square inch |
| 0.0000035274 | ounces per square centimeter |
| 1.0 | grams per square meter |
| 0.092903 | grams per square foot |
| 0.00064521 | grams per square inch |
| 0.0001 | grams per square centimeter |
| 15.4324 | grains per square meter |
| 1.43372 | grains per square foot |
| 0.0099563 | grains per square inch |
| 0.00154324 | grains per square centimeter |

## HECTARE: =

| | |
|---|---|
| 0.003861 | square miles or sections |
| 0.010 | square kilometers |
| 0.247104 | square furlongs |
| 1.0 | square hektometers |
| 24.71044 | square chains |
| 100 | square dekameters |
| 395.367 | square rods |
| 10,000 | square meters |
| 11,959.888 | square yards |
| 13,949.8 | square varas (Texas) |

**HECTARE (cont'd): =**

| | |
|---|---|
| 107,639 | square feet |
| 191,358 | square spans |
| 277,104.4 | square links |
| 968,750 | square hands |
| 1,000,000 | square decimeters |
| 100,000,000 | square centimeters |
| 15,500,016 | square inches |
| 10,000,000,000 | square millimeters |
| 100 | ares |
| 1,000 | centares (centiares) |
| 2.471044 | acres |

**HEKTOLITER: =**

| | |
|---|---|
| 0.000000000053961 | cubic miles |
| 0.0000000001 | cubic kilometers |
| 0.000000000431688 | cubic furlongs |
| 0.0000001 | cubic hektometers |
| 0.0000122835 | cubic chains |
| 0.0001 | cubic dekameters |
| 0.000786142 | cubic rods |
| 0.1 | kiloliters |
| 0.1 | cubic meters |
| 0.13080 | cubic yards |
| 0.164759 | cubic Varas (Texas) |
| 0.628976 | barrels |
| 1.0 | hektoliters |
| 2.8378 | bushels—U.S. (dry) |
| 2.7497 | bushels—Imperial (dry) |
| 3.53145 | cubic feet |
| 8.370844 | cubic spans |
| 10.0 | dekaliters |
| 11.3513 | pecks |
| 12.283475 | cubic links |
| 26.417762 | gallons—U.S. (liquid) |
| 22.702 | gallons—U.S. (dry) |
| 21.998 | gallons—Imperial |
| 95.635866 | cubic hands |
| 100 | liters |
| 100 | cubic decimeters |
| 105.6710 | quarts—U.S. (liquid) |
| 90.8102 | quarts—U.S. (dry) |
| 211.34 | pints—U.S. (liquid) |
| 181.62 | pints—U.S. (dry) |

## HEKTOLITER (cont'd): =

| | |
|---|---|
| 845.38 | gills (liquid) |
| 1,000 | deciliters |
| 61,025 | cubic inches |
| 10,000 | centiliters |
| 100,000 | milliliters |
| 100,000 | cubic centimeters |
| 100,000,000 | cubic millimeters |
| 220.46223 | pounds of water @ 39° F. |
| 3,381.47 | ounces (fluid) |
| 27,051.79 | drams (fluid) |

## HEKTOGRAM: =

| | |
|---|---|
| 0.000098426 | tons (long) |
| 0.0001 | tons (metric) |
| 0.000110231 | tons (short) |
| 0.1 | kilograms |
| 0.220462 | pounds (Troy) |
| 0.267923 | pounds (Avoir.) |
| 1.0 | hektograms |
| 10 | dekagrams |
| 3.21507 | ounces (Troy) |
| 3.527392 | ounces (Avoir.) |
| 25.7206 | drams (Troy) |
| 56.4383 | drams (Avoir.) |
| 64.30149 | pennyweights (Troy) |
| 77.1618 | scruples (Troy) |
| 500 | carats (metric) |
| 1,000 | decigrams |
| 1,543.24 | grains |
| 10,000 | centigrams |
| 100,000 | milligrams |

## HORSEPOWER: =

| | |
|---|---|
| 47,520,000 | foot pounds per day |
| 1,980,000 | foot pounds per hour |
| 33,000 | foot pounds per minute |
| 550 | foot pounds per second |
| 570,240,000 | inch pounds per day |
| 23,760,000 | inch pounds per hour |
| 396,000 | inch pounds per minute |

HORSEPOWER (cont'd): =

| | |
|---|---|
| 6,600 | inch pounds per second |
| 15,390.720 | kilogram calories (mean) per day |
| 641.280 | kilogram calories (mean) per hour |
| 10.688 | kilogram calories (mean) per minute |
| 0.178133 | kilogram calories (mean) per second |
| 33,930.724525 | pounds calories (mean) per day |
| 1,413.780189 | pound calories (mean) per hour |
| 23.563003 | pound calories (mean) per minute |
| 0.392716 | pound calories (mean) per second |
| 542,891.59248 | ounce calories (mean) per day |
| 22,620.483024 | ounce calories (mean) per hour |
| 377.008048 | ounce calories (mean) per minute |
| 6.283456 | ounce calories (mean) per second |
| 15,390,720 | gram calories (mean) per day |
| 641,280 | gram calories (mean) per hour |
| 10,688 | gram calories (mean) per minute |
| 178.133 | gram calories (mean) per second |
| 61,081.344 | BTU (mean) per day |
| 2,545.5600 | BTU (mean) per hour |
| 42.41760 | BTU (mean) per minute |
| 0.70696 | BTU (mean) per second |
| 0.7452 | kilowatts (g=980) |
| 0.74570 | kilowatts (g=980.665) |
| 745.2 | watts (g=980) |
| 745.70 | watts (g=980.665) |
| 1.0139 | horsepower (metric) |
| 1.0139 | Cheval-Vapeur hours |
| 0.174 | pounds carbon oxidized with 100% efficiency |
| 2.62 | pounds water evaporated from and @ 212° F. |
| 635.769600 | kiloliter-atmospheres per day |
| 24.490400 | kiloliter-atmospheres per hour |
| 0.441507 | kiloliter-atmospheres per minute |
| 0.00735844 | kiloliter-atmospheres per second |
| 635,769.599962 | liter-atmospheres per day |
| 26,490.399998 | liter-atmospheres per hour |
| 441.506667 | liter-atmospheres per minute |
| 7.358844 | liter-atmospheres per second |
| 635,769,599.962 | milliliter-atmospheres per day |
| 26,490,399.998 | milliliter-atmospheres per hour |
| 441,506.666667 | milliliter-atmospheres per minute |
| 7,358.444444 | milliliter-atmospheres per second |

## INCH: =

|  |  |
|---|---|
| 0.00001578 | miles |
| 0.00002540 | kilometers |
| 0.000126263 | furlongs |
| 0.0002540 | hektometers |
| 0.00126263 | chains |
| 0.002540 | dekameters |
| 0.00505051 | rods |
| 0.02540 | meters |
| 0.027777 | yards |
| 0.030000 | varas (Texas) |
| 0.083333 | feet |
| 0.111111 | spans |
| 0.126263 | links |
| 0.25000 | hands |
| 0.2540 | decimeters |
| 2.5400 | centimeters |
| 1 | inches |
| 25.40 | millimeters |
| 1000 | mils |
| 25,400 | microns |
| 39,450.33 | wave lengths of red line of cadmium |
| 25,400,000 | millimicrons |
| 25,400,000 | micro-millimeters |
| 254,000,000 | Angstrom Units |

## INCH OF MERCURY @ 32° F. =

|  |  |
|---|---|
| 0.00345349 | hektometers of water @ 60° F. |
| 0.0345349 | dekameters of water @ 60° F. |
| 0.345349 | meters of water @ 60° F. |
| 1.132944 | feet of water @ 60° F. |
| 13.595326 | inches of water @ 60° F. |
| 3.45349 | decimeters of water @ 60° F. |
| 34.5349 | centimeters of water @ 60° F. |
| 345.349 | millimeters of water @ 60° F. |
| 0.000254 | hecktometers of mercury @ 32° F. |
| 0.00254 | dekameters of mercury @ 32° F. |
| 0.0254 | meters of mercury @ 32° F. |
| 0.0833325 | feet of mercury @ 32° F. |
| 1 | inches of mercury @ 32° F. |
| 0.254 | decimeters of mercury @ 32° F. |
| 2.54 | centimeters of mercury @ 32° F. |
| 25.4 | millimeters of mercury @ 32° F. |
| 3806.515240 | tons per square hektometer |

INCH OF MERCURY @ 32° F. (cont'd): =

| | |
|---|---|
| 38.065152 | tons per square dekameter |
| 0.380652 | tons per square meter |
| 0.0353645 | tons per square foot |
| 0.000245581 | tons per square inch |
| 0.00380652 | tons per square decimeter |
| 0.0000380652 | tons per square centimeter |
| 0.000000380652 | tons per square millimeter |
| 3,453,490 | kilograms per square hektometer |
| 34,534.9 | kilograms per square dekameter |
| 345.349 | kilograms per square meter |
| 32.0811278 | kilograms per square foot |
| 0.222786 | kilograms per square inch |
| 3.45349 | kilograms per square decimeter |
| 0.0345349 | kilograms per square centimeter |
| 0.000345349 | kilograms per square millimeter |
| 7,613,030 | pounds per square hektometer |
| 76,130.300609 | pounds per square dekameter |
| 761.303006 | pounds per square meter |
| 70.726441 | pounds per square foot |
| 0.491161 | pounds per square inch |
| 7.613030 | pounds per square decimeter |
| 0.0761303 | pounds per square centimeter |
| 0.000761303 | pounds per square millimeter |
| 34,534,900 | hektograms per square hektometer |
| 345,349 | hektograms per square dekameter |
| 3,453.49 | hektograms per square meter |
| 320.811278 | hektograms per square foot |
| 2.227864 | hektograms per square inch |
| 34.5349 | hektograms per square decimeter |
| 0.345349 | hektograms per square centimeter |
| 0.00345349 | hektograms per square millimeter |
| 345,349,000 | dekagrams per square hektometer |
| 3,453,490 | dekagrams per square dekameter |
| 34,534.90 | dekagrams per square meter |
| 3208.112776 | dekagrams per square foot |
| 22.278639 | dekagrams per square inch |
| 345.3490 | dekagrams per square decimeter |
| 3.453490 | dekagrams per square centimeter |
| 0.03453490 | dekagrams per square millimeter |
| 121,808.481 | ounces per square hektometer |
| 1,218,085 | ounces per square dekameter |
| 12,180.85 | ounces per square meter |
| 1,131.623063 | ounces per square foot |
| 7.858573 | ounces per square inch |

## INCH OF MERCURY @ 32° F. (cont'd): =

121.8085 . . . . . . . . . . . . . . . . . . . . . . . . . ounces per square decimeter
1.218085 . . . . . . . . . . . . . . . . . . . . . . . . .ounces per square centimeter
0.0121809 . . . . . . . . . . . . . . . . . . . . . . . . ounces per square millimeter
3,453,490,000 . . . . . . . . . . . . . . . . . . . .grams per square hektometer
34,534,900 . . . . . . . . . . . . . . . . . . . . . . . grams per square dekameter
345,349 . . . . . . . . . . . . . . . . . . . . . . . . . . . grams per square meter
32,081.127762 . . . . . . . . . . . . . . . . . . . . . . . grams per square foot
222.786665 . . . . . . . . . . . . . . . . . . . . . . . . grams per square inch
3,453.49 . . . . . . . . . . . . . . . . . . . . . . . . .grams per square decimeter
34.5349 . . . . . . . . . . . . . . . . . . . . . . . . . grams per square centimeter
0.345349 . . . . . . . . . . . . . . . . . . . . . . . . grams per square millimeter
34,534,900,000 . . . . . . . . . . . . . . . . . . . decigrams per square hektometer
345,349,000 . . . . . . . . . . . . . . . . . . . . . . . decigrams per square dekameter
3,453,490 . . . . . . . . . . . . . . . . . . . . . . . . . .decigrams per square meter
320.811278 . . . . . . . . . . . . . . . . . . . . . . . . .decigrams per square foot
2,227.866675 . . . . . . . . . . . . . . . . . . . . . . .decigrams per square inch
34,534.90 . . . . . . . . . . . . . . . . . . . . . . . . decigrams per square decimeter
345.3490 . . . . . . . . . . . . . . . . . . . . . . . . . decigrams per square centimeter
3.453490 . . . . . . . . . . . . . . . . . . . . . . . . . decigrams per square millimeter
345,349,000,000 . . . . . . . . . . . . . . . . . . .centigrams per square hektometer
3,453,490,000 . . . . . . . . . . . . . . . . . . . . . centigrams per square dekameter
34,534,900 . . . . . . . . . . . . . . . . . . . . . . . . centigrams per square meter
3,208,113 . . . . . . . . . . . . . . . . . . . . . . . . . . centigrams per square foot
22,278.666751 . . . . . . . . . . . . . . . . . . . . . . centigrams per square inch
345,349 . . . . . . . . . . . . . . . . . . . . . . . . . .centigrams per square decimeter
3,453.49 . . . . . . . . . . . . . . . . . . . . . . . . centigrams per square centimeter
34.5349 . . . . . . . . . . . . . . . . . . . . . . . . . .centigrams per square millimeter
3,453,490,000,000 . . . . . . . . . . . . . . . . . milligrams per square hektometer
34,534,900,000 . . . . . . . . . . . . . . . . . . . .milligrams per square dekameter
345,349,000 . . . . . . . . . . . . . . . . . . . . . . . . milligrams per square meter
32,081,128 . . . . . . . . . . . . . . . . . . . . . . . . . milligrams per square foot
222,786.667776 . . . . . . . . . . . . . . . . . . . . . milligrams per square inch
3,453,490 . . . . . . . . . . . . . . . . . . . . . . . . milligrams per square decimeter
34,534.90 . . . . . . . . . . . . . . . . . . . . . . . .milligrams per square centimeter
345.3490 . . . . . . . . . . . . . . . . . . . . . . . . milligrams per square millimeter
0.0338659 . . . . . . . . . . . . . . . . . . . . . . . . . . . . . . . . . . bars
0.0334214 . . . . . . . . . . . . . . . . . . . . . . . . . . . . . atmospheres
3,383,845,567,678 . . . . . . . . . . . . . . . . . .dynes per square hektometer
33,838,455,677 . . . . . . . . . . . . . . . . . . . . dynes per square dekameter
338,384,557 . . . . . . . . . . . . . . . . . . . . . . . . . dynes per square meter
31,460,290 . . . . . . . . . . . . . . . . . . . . . . . . . . dynes per square foot
218,471.233172 . . . . . . . . . . . . . . . . . . . . . . . dynes per square inch
3,383,846 . . . . . . . . . . . . . . . . . . . . . . . . . .dynes per square decimeter
33,838,455677 . . . . . . . . . . . . . . . . . . . . . . dynes per square centimeter

INCH OF MERCURY @ 32° F. (cont'd): =

338.384557 . . . . . . . . . . . . . . . . . . . . . . . . . dynes per square millimeter
930.464111 . . . . . . . . . . . feet of water @ 62° F. and 29.92 Barom. Press

INCH OF WATER @ 60° F.: =

0.000254 . . . . . . . . . . . . . . . . . . . . . . . . . .hektometers of water @ 60° F.
0.00254 . . . . . . . . . . . . . . . . . . . . . . . . . . dekameters of water @ 60° F.
0.0254 . . . . . . . . . . . . . . . . . . . . . . . . . . . . meters of water @ 60° F.
0.0833332 . . . . . . . . . . . . . . . . . . . . . . . . . .feet of water @ 60° F.
1 . . . . . . . . . . . . . . . . . . . . . . . . . . . . . . inches of water @ 60° F.
0.254 . . . . . . . . . . . . . . . . . . . . . . . . . .decimeters of water @ 60° F.
2.54 . . . . . . . . . . . . . . . . . . . . . . . . . centimeters of water @ 60° F.
25.4 . . . . . . . . . . . . . . . . . . . . . . . . . millimeters of water @ 60° F.
0.0000186820 . . . . . . . . . . . . . . . . . hektometers of mercury @ 32° F.
0.000186820 . . . . . . . . . . . . . . . . . . .dekameters of mercury @ 32° F.
0.00186820 . . . . . . . . . . . . . . . . . . . . . . meters of mercury @ 32° F.
0.00612925 . . . . . . . . . . . . . . . . . . . . . . . feet of mercury @ 32° F.
0.0735510 . . . . . . . . . . . . . . . . . . . . . . .inches of mercury @ 32° F.
0.0186820 . . . . . . . . . . . . . . . . . . . . decimeters of mercury @ 32° F.
0.186820 . . . . . . . . . . . . . . . . . . . . centimeters of mercury @ 32° F.
1.868197 . . . . . . . . . . . . . . . . . . . . .millimeters of mercury @ 32° F.
279.973171 . . . . . . . . . . . . . . . . . . . . . . . . .tons per square hektometer
2.799732 . . . . . . . . . . . . . . . . . . . . . . . . . . tons per square dekameter
0.0279973 . . . . . . . . . . . . . . . . . . . . . . . . . . . . tons per square meter
0.00260110 . . . . . . . . . . . . . . . . . . . . . . . . . . . . tons per square foot
0.0000180627 . . . . . . . . . . . . . . . . . . . . . . . . . . tons per square inch
0.000279973 . . . . . . . . . . . . . . . . . . . . . . . . .tons per square decimeter
0.00000279973 . . . . . . . . . . . . . . . . . . . . . . tons per square centimeter
0.0000000279973 . . . . . . . . . . . . . . . . . . . . . tons per square millimeter
254,000 . . . . . . . . . . . . . . . . . . . . . . . . kilograms per square hektometer
2,540 . . . . . . . . . . . . . . . . . . . . . . . . . . kilograms per square dekameter
25.4 . . . . . . . . . . . . . . . . . . . . . . . . . . . . . .kilograms per square meter
2.359600 . . . . . . . . . . . . . . . . . . . . . . . . . . . .kilograms per square foot
0.0163861 . . . . . . . . . . . . . . . . . . . . . . . . . . .kilograms per square inch
0.254 . . . . . . . . . . . . . . . . . . . . . . . . . . kilograms per square decimeter
0.00254 . . . . . . . . . . . . . . . . . . . . . . . . . kilograms per square centimeter
0.0000254 . . . . . . . . . . . . . . . . . . . . . . . . kilograms per square millimeter
559,946 . . . . . . . . . . . . . . . . . . . . . . . . . . pounds per square hektometer
5,599.463120 . . . . . . . . . . . . . . . . . . . . . . . pounds per square dekameter
55.994631 . . . . . . . . . . . . . . . . . . . . . . . . . . . . pounds per square meter
5.202004 . . . . . . . . . . . . . . . . . . . . . . . . . . . . . pounds per square foot
0.0361250 . . . . . . . . . . . . . . . . . . . . . . . . . . . . pounds per square inch
0.559946 . . . . . . . . . . . . . . . . . . . . . . . . . . . pounds per square decimeter

INCH OF WATER @ 60° F. (cont'd): =

| | |
|---|---|
| 0.00559946 | pounds per square centimeter |
| 0.0000559946 | pounds per square millimeter |
| 2,540,000 | hektograms per square hektometer |
| 25,400 | hektograms per square dekameter |
| 254 | hektograms per square meter |
| 23.596005 | hektograms per square foot |
| 0.163862 | hektograms per square inch |
| 2.54 | hektograms per square decimeter |
| 0.0254 | hektograms per square centimeter |
| 0.000254 | hektograms per square millimeter |
| 25,400,000 | dekagrams per square hektometer |
| 254,000 | dekagrams per square dekameter |
| 2,540 | dekagrams per square meter |
| 235.960045 | dekagrams per square foot |
| 1.638617 | dekagrams per square inch |
| 25.40 | dekagrams per square decimeter |
| 0.2540 | dekagrams per square centimeter |
| 0.002540 | dekagrams per square millimeter |
| 8,959,141 | ounces per square hektometer |
| 89,591.41 | ounces per square dekameter |
| 895.9141 | ounces per square meter |
| 83.232058 | ounces per square foot |
| 0.5780 | ounces per square inch |
| 8.959141 | ounces per square decimeter |
| 0.0895914 | ounces per square centimeter |
| 0.000895914 | ounces per square millimeter |
| 254,000,000 | grams per square hektometer |
| 2,540,000 | grams per square dekameter |
| 25,400 | grams per square meter |
| 2,359.600439 | grams per square foot |
| 16.386192 | grams per square inch |
| 254 | grams per square decimeter |
| 2.54 | grams per square centimeter |
| 0.0254 | grams per square millimeter |
| 2.540,000,000 | decigrams per square hektometer |
| 25,400,000 | decigrams per square dekameter |
| 254,000 | decigrams per square meter |
| 23,596.004 | decigrams per square foot |
| 163.861920 | decigrams per square inch |
| 2,540 | decigrams per square decimeter |
| 25.40 | decigrams per square centimeter |
| 0.2540 | decigrams per square millimeter |
| 25,400,000,000 | centigrams per square hektometer |
| 254,000,000 | centigrams per square dekameter |

**INCH OF WATER @ 60° F. (cont'd): =**

| | |
|---|---|
| 2,540,000 | centigrams per square meter |
| 235,960 | centigrams per square foot |
| 1,638.619198 | centigrams per square inch |
| 25,400 | centigrams per square decimeter |
| 254 | centigrams per square centimeter |
| 2.54 | centigrams per square millimeter |
| 254,000,000,000 | milligrams per square hektometer |
| 2,540,000,000 | milligrams per square dekameter |
| 25,400,000 | milligrams per square meter |
| 2,359,600 | milligrams per square foot |
| 16,386.192 | milligrams per square inch |
| 254,000 | milligrams per square decimeter |
| 2,540 | milligrams per square centimeter |
| 25.40 | milligrams per square millimeter |
| 0.00245562 | bars |
| 0.00245818 | atmospheres |
| 248,885,374,188 | dynes per square hektometer |
| 2,488,853,742 | dynes per square dekameter |
| 24,888,537 | dynes per square meter |
| 2,313,937 | dynes per square foot |
| 16,069.0079 | dynes per square inch |
| 248,885 | dynes per square decimeter |
| 2,488.853742 | dynes per square centimeter |
| 24.888537 | dynes per square millimeter |
| 68.44 | feet of water @ 62° F. and 29.92 Barom. Press. |

**JOULE (ABSOLUTE): =**

| | |
|---|---|
| 23.730 | foot poundals |
| 284.760 | inch poundals |
| 0.73756 | foot pounds |
| 8.85072 | inch pounds |
| 0.000000263331 | ton (net) calories |
| 0.00023889 | kilogram calories (mean) |
| 0.00526661 | pound calories |
| 0.00842658 | ounce calories |
| 0.23889 | gram calories (mean) |
| 238.89 | milligram calories |
| 0.0000000115740 | kilowatt days |
| 0.0000002778 | kilowatt hours |
| 0.0000166667 | kilowatt minutes |
| 0.001 | kilowatt seconds |
| 0.0000115740 | watt days |

JOULE (ABSOLUTE) (cont'd): =

0.0002778 . . . . . . . . . . . . . . . . . . . . . . . . . . . . . watt hours
0.0166667 . . . . . . . . . . . . . . . . . . . . . . . . . . . watt minutes
1 . . . . . . . . . . . . . . . . . . . . . . . . . . . . . . . . . watt seconds
0.000112366 . . . . . . . . . . . . . . . . . . . . . . . . ton (net) meters
0.101937 . . . . . . . . . . . . . . . . . . . . . . . . . kilogram meters
0.224733 . . . . . . . . . . . . . . . . . . . . . . . . . . pounds meters
3.595721 . . . . . . . . . . . . . . . . . . . . . . . . . . . ounce meters
101.937 . . . . . . . . . . . . . . . . . . . . . . . . . . . . gram meters
101,937 . . . . . . . . . . . . . . . . . . . . . . . . . milligram meters
0.000368654 . . . . . . . . . . . . . . . . . . . . . . . . .ton (net) feet
0.334438 . . . . . . . . . . . . . . . . . . . . . . . . . . . kilogram feet
0.737311 . . . . . . . . . . . . . . . . . . . . . . . . . . . . . pound feet
11.796960 . . . . . . . . . . . . . . . . . . . . . . . . . . . . ounce feet
334.438274 . . . . . . . . . . . . . . . . . . . . . . . . . . . . gram feet
334,438.273531 . . . . . . . . . . . . . . . . . . . . . . . milligram feet
0.00442385 . . . . . . . . . . . . . . . . . . . . . . . ton (net) inches
4.013259 . . . . . . . . . . . . . . . . . . . . . . . . . kilogram inches
8.847732 . . . . . . . . . . . . . . . . . . . . . . . . . . . pound inches
141.563520 . . . . . . . . . . . . . . . . . . . . . . . . . . ounce inches
4,013.259288 . . . . . . . . . . . . . . . . . . . . . . . . . .gram inches
4,013,259 . . . . . . . . . . . . . . . . . . . . . . . . . milligram inches
0.0112366 . . . . . . . . . . . . . . . . . . . . . . ton (net) centimeters
10.1937 . . . . . . . . . . . . . . . . . . . . . . . . kilogram centimeters
22.4733 . . . . . . . . . . . . . . . . . . . . . . . . . .pound centimeters
359.5721 . . . . . . . . . . . . . . . . . . . . . . . . . ounce centimeters
10,193.7 . . . . . . . . . . . . . . . . . . . . . . . . . . .gram centimeters
10,193,700 . . . . . . . . . . . . . . . . . . . . . . . milligram centimeters
0.112366 . . . . . . . . . . . . . . . . . . . . . . . . ton (net) millimeters
101.937 . . . . . . . . . . . . . . . . . . . . . . . . .kilogram millimeters
224.733 . . . . . . . . . . . . . . . . . . . . . . . . . . pound millimeters
3,595.721 . . . . . . . . . . . . . . . . . . . . . . . . . ounce millimeters
101,937 . . . . . . . . . . . . . . . . . . . . . . . . . . . gram millimeters
101,937,000 . . . . . . . . . . . . . . . . . . . . . . milligram millimeters
0.0000098705 . . . . . . . . . . . . . . . . . . . . . kiloliter-atmospheres
0.000098705 . . . . . . . . . . . . . . . . . . . . hektoliter-atmospheres
0.0003485 . . . . . . . . . . . . . . . . . . . . . . cubic foot-atmospheres
0.00098705 . . . . . . . . . . . . . . . . . . . . . . dekaliter-atmospheres
0.0098705 . . . . . . . . . . . . . . . . . . . . . . . . . .liter-atmospheres
0.098705 . . . . . . . . . . . . . . . . . . . . . . . . deciliter-atmospheres
0.98705 . . . . . . . . . . . . . . . . . . . . . . . . .centiliter-atmospheres
9.8705 . . . . . . . . . . . . . . . . . . . . . . . . millimeter-atmospheres
1 . . . . . . . . . . . . . . . . . . . . . . . . . . . . . . . . . . . . . . joules
0.0000003775 . . . . . . . . . . . . . . . . . . . . . . Cheval-Vapeur hours
0.0000000155208 . . . . . . . . . . . . . . . . . . . . . . . horsepower days

## JOULE (ABSOLUTE) (cont'd): =

| | |
|---|---|
| 0.0000003725 | horsepower hours |
| 0.0000223500 | horsepower minutes |
| 0.00134100 | horsepower seconds |
| 0.0000000642 | pounds of carbon oxidized with perfect efficiency |
| 0.0000009662 | pounds of water evaporated from and at 212° F. |
| 0.0009480 | BTU (mean) |
| 100,000,000 | ergs |

## KILOGRAMS: =

| | |
|---|---|
| 0.000984206 | tons (long) |
| 0.001 | tons (metric) |
| 0.00110231 | tons (short) |
| 1 | kilograms |
| 2.679229 | pounds (Troy) |
| 2.204622 | pounds (Avoir.) |
| 10 | hektograms |
| 100 | dekagrams |
| 32.150742 | ounces (Troy) |
| 35.273957 | ounces (Avoir.) |
| 1,000 | grams |
| 257.21 | drams (Troy) |
| 564.38 | drams (Avoir.) |
| 643.01 | pennyweights (Troy) |
| 771.62 | scruples (Troy) |
| 5,000 | carats (metric) |
| 10,000 | decigrams |
| 15,432.4 | grains |
| 100,000 | centigrams |
| 1,000,000 | milligrams |

## KILOGRAM CALORIE (MEAN): =

| | |
|---|---|
| 99,334 | foot poundals |
| 1,192,008 | inch poundals |
| 3,087.4 | foot pounds |
| 37,048.8 | inch pounds |
| 0.0001 | ton (short) calories |
| 1 | kilogram calories |
| 2.04622 | pound calories |
| 10 | hektogram calories |
| 100 | dekagram calories |

## KILOGRAM CALORIES (MEAN) (cont'd): =

| | |
|---|---|
| 32.7395 | ounce calories |
| 1,000 | gram calories |
| 10,000 | decigram calories |
| 100,000 | centigram calories |
| 1,000,000 | milligram calories |
| 0.00004845 | kilowatt days |
| 0.0011628 | kilowatt hours |
| 0.0697680 | kilowatt minutes |
| 4.186 | kilowatt seconds |
| 0.04845 | watt days |
| 1.1628 | watt hours |
| 69.7680 | watt minutes |
| 4,186 | watt seconds |
| 0.42685 | ton meters |
| 426.85 | kilogram meters |
| 941.043 | pound meters |
| 4,268.5 | hektogram meters |
| 42,685 | dekagram meters |
| 15,056.688 | ounce meters |
| 426,850 | gram meters |
| 4,268,500 | decigram meters |
| 42,685,000 | centigram meters |
| 426,850,000 | milligram meters |
| 0.0042685 | ton hektometers |
| 4.2685 | kilogram hektometers |
| 9.41043 | pound hektometers |
| 42.685 | hektogram hektometers |
| 426.85 | dekagram hektometers |
| 150.567 | ounce hektometers |
| 4,268.5 | gram hektometers |
| 42,685 | decigram hektometers |
| 426,850 | centigram hektometers |
| 4,268,500 | milligram hektometers |
| 0.042685 | ton dekameters |
| 42.685 | kilogram dekameters |
| 94.1043 | pound dekameters |
| 426.85 | hektogram dekameters |
| 4,268.5 | dekagram dekameters |
| 1,505.67 | ounce dekameters |
| 42,685 | gram dekameters |
| 426,850 | decigram dekameters |
| 4,268,500 | centigram dekameters |
| 42,685,000 | milligram dekameters |
| 1.40042 | ton feet |

## KILOGRAM CALORIE (MEAN) (cont'd): =

| | |
|---|---|
| 1,400.424 | kilogram feet |
| 30.8704 | pound feet |
| 14,004.236 | hektogram feet |
| 140,042.357 | dekagram feet |
| 493.985 | ounce feet |
| 1,400,424 | gram feet |
| 14,004,236 | decigram feet |
| 140,042,357 | centigram feet |
| 1,400,423,566 | milligram feet |
| 16.8050 | ton inches |
| 16,805.083 | kilogram inches |
| 370.445 | pound inches |
| 168,050.828 | hektogram inches |
| 1,680,051 | dekagram inches |
| 5,927.117 | ounce inches |
| 16,805,083 | gram inches |
| 168,050,828 | decigram inches |
| 1,680,508,279 | centigram inches |
| 16,805,082,792 | milligram inches |
| 4.2685 | ton decimeters |
| 4,268.5 | kilogram decimeters |
| 9,410.430 | pound decimeters |
| 42,685 | hektogram decimeters |
| 426,850 | dekogram decimeters |
| 150,566.885 | ounce decimeters |
| 4,268,500 | gram decimeters |
| 42,685,000 | decigram decimeters |
| 426,850,000 | centigram decimeters |
| 4,268,500,000 | milligram decimeters |
| 42.685 | ton centimeters |
| 42,685 | kilogram centimeters |
| 94,104.303 | pound centimeters |
| 426,850 | hektogram centimeters |
| 4,268,500 | dekagram centimeters |
| 1,505,669 | ounce centimeters |
| 42,685,000 | gram centimeters |
| 426,850,000 | decigram centimeters |
| 4,268,500,000 | centigram centimeters |
| 42,685,000,000 | milligram centimeters |
| 426.85 | ton millimeters |
| 426,850 | kilogram millimeters |
| 941,043.029 | pound millimeters |
| 4,268,500 | hektogram millimeters |
| 42,685,000 | dekagram millimeters |

## KILOGRAM CALORIE (MEAN) (cont'd): =

| | |
|---|---|
| 15,056,688 | ounce millimeters |
| 426,850,000 | gram millimeters |
| 4,268,500,000 | decigram millimeters |
| 42,685,000,000 | centigram millimeters |
| 426,850,000,000 | milligram millimeters |
| 0.041311 | kiloliter-atmospheres |
| 0.41311 | hektoliter-atmospheres |
| 4.1311 | dekaliter-atmospheres |
| 41.311 | liter-atmospheres |
| 413.11 | deciliter-atmospheres |
| 4,131.1 | centiliter-atmospheres |
| 41,311 | milliliter-atmospheres |
| 0.0000000000415977 | cubic kilometr-atmospheres |
| 0.0000000415977 | cubic hektometer-atmospheres |
| 0.0000415977 | cubic dekameter-atmospheres |
| 0.0415977 | cubic meter-atmospheres |
| 1.469 | cubic feet-atmospheres |
| 2,538.415 | cubic inch-atmospheres |
| 41.5977 | cubic decimeter-atmospheres |
| 41,597.673 | cubic centimeter-atmospheres |
| 41,597,673 | cubic millimeter-atmospheres |
| 4,185.8291 | joules |
| 0.00158097 | Cheval-Vapeur hours |
| 0.0000649708 | horsepower days |
| 0.0015593 | horsepower hours |
| 0.0935980 | horsepower minutes |
| 5.61588 | horsepower seconds |
| 0.00029909 | pounds of carbon oxidized with 100% efficiency |
| 0.004501 | pounds of water evaporated from at 212° F. |
| 1.800 | BTU (mean) per pound |
| 3.9685 | BTU (mean) |
| 41,858,291,000 | ergs |

## KILOGRAM CALORIE (MEAN) PER MINUTE: =

| | |
|---|---|
| 4,443,725 | foot pound per day |
| 185,155.2 | foot pound per hour |
| 3,085.920 | foot pound per minute |
| 51.432 | foot pound per second |
| 0.0935980 | horsepower |
| 0.069680 | kilowatts |
| 69.7680 | watts |

## KILOGRAM METER: =

|  |  |
|---|---|
| 232.71 | foot poundals |
| 2,792.52 | inch poundals |
| 7.2330 | foot pounds |
| 86.7960 | inch pounds |
| 0.0000023427 | ton (net) calories |
| 0.0023427 | kilogram calories |
| 0.00516477 | pound calories |
| 0.023427 | hektogram calories |
| 0.23427 | dekagram calories |
| 0.0826363 | ounce calories |
| 2.3427 | gram calories (mean) |
| 23.427 | decigram calories |
| 234.27 | centigram calories |
| 2,342.7 | milligram calories |
| 0.000000113479 | kilowatt days |
| 0.0000027235 | kilowatt hours |
| 0.000163410 | kilowatt minutes |
| 0.00980460 | kilowatt seconds |
| 0.000113479 | watt days |
| 0.0027235 | watt hours |
| 0.163410 | watt minutes |
| 9.80460 | watt seconds |
| 0.001 | ton meters |
| 1 | kilogram meters |
| 2.204622 | pound meters |
| 10 | hektogram meters |
| 100 | dekagram meters |
| 35.273957 | ounce meters |
| 1,000 | gram meters |
| 10,000 | decigram meters |
| 100,000 | centigram meters |
| 1,000,000 | milligram meters |
| 0.00001 | ton hektometers |
| 0.01 | kilogram hektometers |
| 0.02204622 | pound hektometers |
| 0.1 | hektogram hektometers |
| 1 | dekagram hektometers |
| 0.352740 | ounce hektometers |
| 10 | gram hektometers |
| 100 | decigram hektometers |
| 1,000 | centigram hektometers |
| 10,000 | milligram hektometers |
| 0.0001 | ton dekameters |
| 0.1 | kilogram dekameters |

KILOGRAM METER (cont'd): =

| | |
|---|---|
| 0.2204622 | pound dekameters |
| 1 | hektogram dekameters |
| 10 | dekagram dekameters |
| 3.527396 | ounce dekameters |
| 100 | gram dekameters |
| 1,000 | decigram dekameters |
| 10,000 | centigram dekameters |
| 100,000 | milligram dekameters |
| 0.00328084 | ton feet |
| 3.280843 | kilogram feet |
| 7.233020 | pound feet |
| 32.80843 | hektogram feet |
| 328.0843 | dekagram feet |
| 115.728320 | ounce feet |
| 3,280.843 | gram feet |
| 32,808.43 | decigram feet |
| 328,084.3 | centigram feet |
| 3,280,843 | milligram feet |
| 0.0393701 | ton inches |
| 39.370116 | kilogram inches |
| 86.796236 | pound inches |
| 393.70116 | hektogram inches |
| 3,937.0116 | dekagram inches |
| 1,388.739776 | ounce inches |
| 39,370.116 | gram inches |
| 393,701.16 | decigram inches |
| 3,937,012 | centigram inches |
| 39,370.116 | milligram inches |
| 0.01 | ton decimeters |
| 10 | kilogram decimeters |
| 22.046223 | pound decimeters |
| 100 | hektogram decimeters |
| 1,000 | dekagram decimeters |
| 352.739568 | ounce decimeters |
| 10,000 | gram decimeters |
| 100,000 | decigram decimeters |
| 1,000,000 | centigram decimeters |
| 10,000,000 | milligram decimeters |
| 0.1 | ton centimeters |
| 100 | kilogram centimeters |
| 220.46223 | pound centimeters |
| 1,000 | hektogram centimeters |
| 10,000 | dekagram centimeters |
| 3,527.39568 | ounce centimeters |

KILOGRAM METER (cont'd): =

100,000 . . . . . . . . . . . . . . . . gram centimeters
1,000,000 . . . . . . . . . . . . . . . . decigram centimeters
10,000,000 . . . . . . . . . . . . . . . . centigram centimeters
100,000,000 . . . . . . . . . . . . . . . . milligram centimeters
1 . . . . . . . . . . . . . . . . ton millimeters
1,000 . . . . . . . . . . . . . . . . kilogram millimeters
2,204.6223 . . . . . . . . . . . . . . . . pound millimeters
10,000 . . . . . . . . . . . . . . . . hektogram millimeters
100,000 . . . . . . . . . . . . . . . . dekagram millimeters
35,273.9568 . . . . . . . . . . . . . . . . ounce millimeters
1,000,000 . . . . . . . . . . . . . . . . gram millimeters
10,000,000 . . . . . . . . . . . . . . . . decigram millimeters
100,000,000 . . . . . . . . . . . . . . . . centigram millimeters
1,000,000,000 . . . . . . . . . . . . . . . . milligram millimeters
0.000096782 . . . . . . . . . . . . . . . . kiloliter-atmospheres
0.00096782 . . . . . . . . . . . . . . . . hektoliter-atmospheres
0.0096782 . . . . . . . . . . . . . . . . dekaliter-atmospheres
0.096782 . . . . . . . . . . . . . . . . liter-atmospheres
0.96782 . . . . . . . . . . . . . . . . deciliter-atmospheres
9.6782 . . . . . . . . . . . . . . . . centiliter-atmospheres
96.782 . . . . . . . . . . . . . . . . milliliter-atmospheres
0.0000000000000967790 . . . . . . . . . . . . . . . . cubic kilometer-atmospheres
0.0000000000967790 . . . . . . . . . . . . . . . . cubic hektometer-atmospheres
0.0000000967790 . . . . . . . . . . . . . . . . cubic dekameter-atmospheres
0.0000967790 . . . . . . . . . . . . . . . . cubic meter-atmospheres
0.0034177 . . . . . . . . . . . . . . . . cubic foot-atmospheres
5.905746 . . . . . . . . . . . . . . . . cubic inch-atmospheres
0.0967790 . . . . . . . . . . . . . . . . cubic decimeter-atmospheres
96.779011 . . . . . . . . . . . . . . . . cubic centimeter-atmospheres
96,779.011 . . . . . . . . . . . . . . . . cubic millimeter-atmospheres
9.80665 . . . . . . . . . . . . . . . . joules
0.000000154324 . . . . . . . . . . . . . . . . Cheval-Vapeur hours
0.000000152208 . . . . . . . . . . . . . . . . horsepower days
0.0000036530 . . . . . . . . . . . . . . . . horsepower hours
0.000219180 . . . . . . . . . . . . . . . . horsepower minutes
0.0131508 . . . . . . . . . . . . . . . . horsepower seconds
0.00000063718 . . . . . . . pounds of carbon oxidized with 100% efficiency
0.0000095895 . . . . . . . pounds of water evaporated from and at $212^\circ$ F.
4.216948 . . . . . . . . . . . . . . . . BTU (mean) per pound
0.0092972 . . . . . . . . . . . . . . . . BTU (mean)
98,066,500 . . . . . . . . . . . . . . . . ergs

## KILOGRAM METER PER SECOND: =

0.0098046 . . . . . . . . . . kilowatts
9.8046 . . . . . . . . . . watts
0.0131508 . . . . . . . . . . horsepower

## KILOGRAM METER PER MINUTE: =

0.000163410 . . . . . . . . . . kilowatts
0.163410 . . . . . . . . . . watts
0.000219180 . . . . . . . . . . horsepowers

## KILOGRAM PER METER: =

1.774004 . . . . . . . . . . tons (net) per mile
1.102311 . . . . . . . . . . tons (net) per kilometer
0.00110231 . . . . . . . . . . tons (net) per meter
0.001007956 . . . . . . . . . . tons (net) per yard
0.000335985 . . . . . . . . . . tons (net) per foot
0.0000279987 . . . . . . . . . . tons (net) per inch
0.0000110231 . . . . . . . . . . tons (net) per centimeter
0.00000110231 . . . . . . . . . . tons (net) per millimeter
1,609.349954 . . . . . . . . . . kilograms per mile
1,000 . . . . . . . . . . kilograms per kilometer
1 . . . . . . . . . . kilograms per meter
0.914403 . . . . . . . . . . kilograms per yard
0.304801 . . . . . . . . . . kilograms per foot
0.0254001 . . . . . . . . . . kilograms per inch
0.01 . . . . . . . . . . kilograms per centimeter
0.001 . . . . . . . . . . kilograms per millimeter
3,548.00896449 . . . . . . . . . . pounds (Avoir.) per mile
2,204.622341 . . . . . . . . . . pounds (Avoir.) per kilometer
2.204622 . . . . . . . . . . pounds (Avoir.) per meter
2.0159127 . . . . . . . . . . pounds (Avoir.) per yard
0.671971 . . . . . . . . . . pounds (Avoir.) per foot
0.0559976 . . . . . . . . . . pounds (Avoir.) per inch
0.0220462 . . . . . . . . . . pounds (Avoir.) per centimeter
0.00220462 . . . . . . . . . . pounds (Avoir.) per millimeter
56,768.143440 . . . . . . . . . . ounces (Avoir.) per mile
35,273.957456 . . . . . . . . . . ounces (Avoir.) per kilometer
35.273957 . . . . . . . . . . ounces (Avoir.) per meter
32.254604 . . . . . . . . . . ounces (Avoir.) per yard
10.751535 . . . . . . . . . . ounces (Avoir.) per foot
0.895961 . . . . . . . . . . ounces (Avoir.) per inch

**KILOGRAM PER METER (cont'd): =**

| | |
|---|---|
| 0.352740 | ounces (Avoir.) per centimeter |
| 0.0352740 | ounces (Avoir.) per millimeter |
| 1,609,349.954 | grams per mile |
| 1,000,000 | grams per kilometer |
| 1,000 | grams per meter |
| 914.403 | grams per yard |
| 304.801127 | grams per foot |
| 25.4001 | grams per inch |
| 10 | grams per centimeter |
| 1 | grams per millimeter |
| 24,836,063 | grains per mile |
| 15,432,356 | grains per kilometer |
| 15,432.356387 | grains per meter |
| 14,111.388900 | grains per yard |
| 4,703.79630 | grains per foot |
| 391.983025 | grains per inch |
| 154.323564 | grains per centimeter |
| 15.432356 | grains per millimeter |

**KILOGRAM PER SQUARE CENTIMETER: =**

| | |
|---|---|
| 10 | meters of water @ 60° F. |
| 32.80833 | feet of water @ 60° F. |
| 393.69996 | inches of water @ 60° F. |
| 1,000 | centimeters of water @ 60° F. |
| 10,000 | millimeters of water @ 60° F. |
| 0.735499 | meters of mercury @ 32° F. |
| 2.413053 | feet of mercury @ 32° F. |
| 28.956632 | inches of mercury @ 32° F. |
| 73.54985 | centimeters of mercury @ 32° F. |
| 735.49845 | millimeters of mercury @ 32° F. |
| 10,000 | kilograms per square meter |
| 929.034238 | kilograms per square foot |
| 6.451626 | kilograms per square inch |
| 1.0 | kilograms per square centimeter |
| 0.01 | kilograms per square millimeter |
| 22,046.223 | pounds per square meter |
| 2,048.1696 | pounds per square foot |
| 14.2234 | pounds per square inch |
| 2.2046223 | pounds per square centimeter |
| 0.0220462 | pounds per square millimeter |
| 352,739.568 | ounces per square meter |
| 32,770.7136 | ounces per square foot |

## KILOGRAM PER SQUARE CENTIMETER (cont'd): =

| | |
|---|---|
| 227.5744 | ounces per square inch |
| 35.273957 | ounces per square centimeter |
| 0.352740 | ounces per square millimeter |
| 10,000,000 | grams per square meter |
| 929,034.230 | grams per square foot |
| 6,451.626597 | grams per square inch |
| 1,000 | grams per square centimeter |
| 10 | grams per square millimeter |
| 154,324,000 | grains per square meter |
| 14,337,228 | grains per square foot |
| 99,564.0822955 | grains per square inch |
| 15,432.4 | grains per square centimeter |
| 154.324 | grains per square millimeter |
| 0.967778 | atmospheres |

## KILOGRAM PER SQUARE FOOT: =

| | |
|---|---|
| 0.0107638 | meters of water @ 60° F. |
| 0.0353518 | feet of water @ 60° F. |
| 0.423774 | inches of water @ 60° F. |
| 1.076387 | centimeters of water @ 60° F. |
| 10.76387 | millimeters of water @ 60° F. |
| 0.000791682 | meters of mercury @ 32° F. |
| 0.00259738 | feet of mercury @ 32° F. |
| 0.0311689 | inches of mercury @ 32° F. |
| 0.0791682 | centimeters of mercury @ 32° F. |
| 0.791682 | millimeters of mercury @ 32° F. |
| 10.76387 | kilograms per square meter |
| 1 | kilograms per square foot |
| 0.00694445 | kilograms per square inch |
| 0.00107639 | kilograms per square centimeter |
| 0.0000107639 | kilograms per square millimeter |
| 23.730265 | pounds per square meter |
| 2.204622 | pounds per square foot |
| 0.0153099 | pounds per square inch |
| 0.00237302 | pounds per square centimeter |
| 0.0000237302 | pounds per square millimeter |
| 379.684288 | ounces per square meter |
| 35.273957 | ounces per square foot |
| 0.244958 | ounces per square inch |
| 0.0379685 | ounces per square centimeter |
| 0.000379685 | ounces per square millimeter |
| 10,763.87 | grams per square meter |

## KILOGRAM PER SQUARE FOOT (cont'd): =

1,000 . . . . . . . . . . . . . . . . . . . . . . . . . . . . . grams per square foot
6.94445 . . . . . . . . . . . . . . . . . . . . . . . . . . . . grams per square inch
1.076387 . . . . . . . . . . . . . . . . . . . . . . . . grams per square centimeter
0.0107639 . . . . . . . . . . . . . . . . . . . . . . . . grams per square millimeter
166,112.347388 . . . . . . . . . . . . . . . . . . . . . . grains per square meter
15,432.4 . . . . . . . . . . . . . . . . . . . . . . . . . . . grains per square foot
107.169481 . . . . . . . . . . . . . . . . . . . . . . . . . grains per square inch
16.611235 . . . . . . . . . . . . . . . . . . . . . . . grains per square centimeter
0.166112 . . . . . . . . . . . . . . . . . . . . . . . . . grains per square millimeter
0.00104170 . . . . . . . . . . . . . . . . . . . . . . . . . . . . . . . atmospheres

## KILOGRAM PER SQUARE INCH: =

1.549987 . . . . . . . . . . . . . . . . . . . . . . . . . . meters of water @ 60° F.
5.090659 . . . . . . . . . . . . . . . . . . . . . . . . . . .feet of water @ 60° F.
61.023456 . . . . . . . . . . . . . . . . . . . . . . . . inches of water @ 60° F.
154.999728 . . . . . . . . . . . . . . . . . . . . . . centimeters of water @ 60° F.
1,549.99728 . . . . . . . . . . . . . . . . . . . . . millimeters of water @ 60° F.
0.114002 . . . . . . . . . . . . . . . . . . . . . . . . . meters of mercury @ 32° F.
0.374022 . . . . . . . . . . . . . . . . . . . . . . . . . . . feet of mercury @ 32° F.
4.488322 . . . . . . . . . . . . . . . . . . . . . . . . . .inches of mercury @ 32° F.
11.40002 . . . . . . . . . . . . . . . . . . . . . . centimeters of mercury @ 32° F.
114.00022 . . . . . . . . . . . . . . . . . . . . . . .millimeters of mercury @ 32° F.
1,549.99728 . . . . . . . . . . . . . . . . . . . . . . . kilograms per square meter
144 . . . . . . . . . . . . . . . . . . . . . . . . . . . . . .kilograms per square foot
1 . . . . . . . . . . . . . . . . . . . . . . . . . . . . . . .kilograms per square inch
0.1549997 . . . . . . . . . . . . . . . . . . . . . . kilograms per square centimeter
0.001549997 . . . . . . . . . . . . . . . . . . . . . kilograms per square millimeter
3,417.158160 . . . . . . . . . . . . . . . . . . . . . . . . . pounds per square meter
317.465568 . . . . . . . . . . . . . . . . . . . . . . . . . . pounds per square foot
2.2046223 . . . . . . . . . . . . . . . . . . . . . . . . . . . pounds per square inch
0.341715 . . . . . . . . . . . . . . . . . . . . . . . . pounds per square centimeter
0.00341715 . . . . . . . . . . . . . . . . . . . . . . . .pounds per square millimeter
54,674.537472 . . . . . . . . . . . . . . . . . . . . . . . . ounces per square meter
5,079.449808 . . . . . . . . . . . . . . . . . . . . . . . . . ounces per square foot
35.273957 . . . . . . . . . . . . . . . . . . . . . . . . . . . ounces per square inch
5.467464 . . . . . . . . . . . . . . . . . . . . . . . . .ounces per square centimeter
0.0546746 . . . . . . . . . . . . . . . . . . . . . . . .ounces per square millimeter
1,549,997 . . . . . . . . . . . . . . . . . . . . . . . . . . . . grams per square meter
144,000 . . . . . . . . . . . . . . . . . . . . . . . . . . . . . . grams per square foot
1,000 . . . . . . . . . . . . . . . . . . . . . . . . . . . . . . . grams per square inch
154.999728 . . . . . . . . . . . . . . . . . . . . . . . . grams per square centimeter
1.549997 . . . . . . . . . . . . . . . . . . . . . . . . . . grams per square millimeter

KILOGRAM PER SQUARE INCH (cont'd): =

| | |
|---|---|
| 23,920,178 | grains per square meter |
| 2,222,266 | grains per square foot |
| 15,432.4 | grains per square inch |
| 2,392.017840 | grains per square centimeter |
| 23.920178 | grains per square millimeter |
| 0.150005 | atmospheres |

KILOGRAM PER SQUARE KILOMETER: =

| | |
|---|---|
| 0.000000001 | meters of water @ 60° F. |
| 0.00000032843 | feet of water @ 60° F. |
| 0.00000003937 | inches of water @ 60° F. |
| 0.0000001 | centimeters of water @ 60° F. |
| 0.000001 | millimeters of water @ 60° F. |
| 0.0000000000735499 | meters of mercury @ 32° F. |
| 0.000000000241305 | feet of mercury @ 32° F. |
| 0.00000000289570 | inches of mercury @ 32° F. |
| 0.00000000735499 | centimeters of mercury @ 32° F. |
| 0.0000000735499 | millimeters of mercury @ 32° F. |
| 0.000001 | kilograms per square meter |
| 0.0000000929034 | kilograms per square foot |
| 0.000000000645163 | kilograms per square inch |
| 0.0000000001 | kilograms per square centimeter |
| 0.000000000001 | kilograms per square millimeter |
| 0.00000220462 | pounds per square meter |
| 0.000000204817 | pounds per square foot |
| 0.00000000142234 | pounds per square inch |
| 0.000000000220462 | pounds per square centimeter |
| 0.00000000000220462 | pounds per square millimeter |
| 0.0000352740 | ounces per square meter |
| 0.00000327707 | ounces per square foot |
| 0.0000000227574 | ounces per square inch |
| 0.00000000352740 | ounces per square centimeter |
| 0.0000000000352740 | ounces per square millimeter |
| 0.001 | grams per square meter |
| 0.0000929034 | grams per square foot |
| 0.000000645163 | grams per square inch |
| 0.0000001 | grams per square centimeter |
| 0.000000001 | grams per square millimeter |
| 0.0154324 | grains per square meter |
| 0.00143372 | grains per square foot |
| 0.00000995640 | grains per square inch |
| 0.00000154324 | grains per square centimeter |

KILOGRAM PER SQUARE KILOMETER (cont'd): =

| | |
|---|---|
| 0.0000000154324 | grains per square millimeter |
| 0.0000000000967778 | atmospheres |

KILOGRAM PER SQUARE METER: =

| | |
|---|---|
| 0.001 | meters of water @ 60° F. |
| 0.0032843 | feet of water @ 60° F. |
| 0.03937 | inches of water @ 60° F. |
| 0.1 | centimeters of water @ 60° F. |
| 1 | millimeters of water @ 60° F. |
| 0.0000735499 | meters of mercury @ 32° F. |
| 0.000241300 | feet of mercury @ 32° F. |
| 0.00289570 | inches of mercury @ 32° F. |
| 0.0073549 | centimeters of mercury @ 32° F. |
| 0.0735499 | millimeters of mercury @ 32° F. |
| 1 | kilograms per square meter |
| 0.0929034 | kilograms per square foot |
| 0.000645163 | kilograms per square inch |
| 0.0001 | kilograms per square centimeter |
| 0.000001 | kilograms per square millimeter |
| 2.204622 | pounds per square meter |
| 0.204817 | pounds per square foot |
| 0.00142234 | pounds per square inch |
| 0.000220462 | pounds per square centimeter |
| 0.00000220462 | pounds per square millimeter |
| 35.273957 | ounces per square meter |
| 3.277071 | ounces per square foot |
| 0.0227574 | ounces per square inch |
| 0.00352740 | ounces per square centimeter |
| 0.0000352740 | ounces per square millimeter |
| 1,000 | grams per square meter |
| 92.903423 | grams per square foot |
| 0.645163 | grams per square inch |
| 0.1 | grams per square centimeter |
| 0.001 | grams per square millimeter |
| 15,432.4 | grains per square meter |
| 1,433.7228 | grains per square foot |
| 9,956408 | grains per square inch |
| 1.543240 | grains per square centimeter |
| 0.0154324 | grains per square millimeter |
| 0.0000967778 | atmospheres |

## KILOGRAM PER SQUARE MILLIMETER: =

| | |
|---|---|
| 1,000 | meters of water @ 60° F. |
| 3,280.833 | feet of water @ 60° F. |
| 39,369.996 | inches of water @ 60° F. |
| 100,000 | centimeters of water @ 60° F. |
| 1,000,000 | millimeters of water @ 60° F. |
| 73.549845 | meters of mercury @ 32° F. |
| 241.3053 | feet of mercury @ 32° F. |
| 2,895.6632 | inches of mercury @ 32° F. |
| 7,354.9845 | centimeters of mercury @ 32° F. |
| 73,549.845 | millimeters of mercury @ 32° F. |
| 1,000,000 | kilograms per square meter |
| 92,903.4238 | kilograms per square foot |
| 645.1626 | kilograms per square inch |
| 100 | kilograms per square centimeter |
| 1 | kilograms per square millimeter |
| 2,204,622 | pounds per square meter |
| 204,816.6 | pounds per square foot |
| 1,422.34 | pounds per square inch |
| 220.46223 | pounds per square centimeter |
| 2.204622 | pounds per square millimeter |
| 35,273,957 | ounces per square meter |
| 3,277,071 | ounces per square foot |
| 22,757.44 | ounces per square inch |
| 3,527.3957 | ounces per square centimeter |
| 35.273957 | ounces per square millimeter |
| 1,000,000,000 | grams per square meter |
| 92,903,423 | grams per square foot |
| 645,162.6597 | grams per square inch |
| 100,000 | grams per square centimeter |
| 1,000 | grams per square millimeter |
| 15,432,400,000 | grains per square meter |
| 1,433,722,800 | grains per square foot |
| 9,956,408 | grains per square inch |
| 1,543,240 | grains per square centimeter |
| 15,432.4 | grains per square millimeter |
| 96.7778 | atmospheres |

## KILOMETER: =

| | |
|---|---|
| 0.53961 | miles (nautical) |
| 0.62137 | miles (statute) |
| 1 | kilometers |
| 4.970974 | furlongs |

**KILOMETER (cont'd): =**

| | |
|---|---|
| 10 | hektometers |
| 49.709741 | chains |
| 100 | dekameters |
| 198.838579 | rods |
| 1,000 | meters |
| 1,093.6 | yards |
| 1,181.1 | varas (Texas) |
| 3,280.8 | feet |
| 4,374.440070 | spans |
| 4,970.974310 | links |
| 9,842.50 | hands |
| 10,000 | decimeters |
| 100,000 | centimeters |
| 39,370 | inches |
| 1,000,000 | millimeters |
| 39,370,000 | mils |
| 1,000,000,000 | microns |
| 1,000,000,000,000 | milli-microns |
| 1,000,000,000,000 | micro-millimeters |
| 546.81 | fathoms |

**KILOMETER PER DAY: =**

| | |
|---|---|
| 0.62137 | miles per day |
| 0.0258904 | miles per hour |
| 0.000431507 | miles per minute |
| 0.00000719718 | miles per second |
| 1.0 | kilometers per day |
| 0.0416667 | kilometers per hour |
| 0.000694444 | kilometers per minute |
| 0.0000115741 | kilometers per second |
| 4.970974 | furlongs per day |
| 0.207124 | furlongs per hour |
| 0.00345207 | furlongs per minute |
| 0.0000575344 | furlongs per second |
| 10 | hektometers per day |
| 0.416667 | hektometers per hour |
| 0.00694444 | hektometers per minute |
| 0.00115741 | hektometers per second |
| 49.709741 | chains per day |
| 2.071239 | chains per hour |
| 0.0345207 | chains per minute |
| 0.00057534 | chains per second |

KILOMETER PER DAY (cont'd): =

| | |
|---|---|
| 100 | dekameters per day |
| 4.166667 | dekameters per hour |
| 0.0694444 | dekameters per minute |
| 0.00115741 | dekameters per second |
| 198.838579 | rods per day |
| 8.284941 | rods per hour |
| 0.138082 | rods per minute |
| 0.00230137 | rods per second |
| 1,000 | meters per day |
| 41.666667 | meters per hour |
| 0.694444 | meters per minute |
| 0.0115741 | meters per second |
| 1,093.6 | yards per day |
| 45.566667 | yards per hour |
| 0.759444 | yards per minute |
| 0.0126574 | yards per second |
| 1,181.1 | varas (Texas) per day |
| 49.21250 | varas (Texas) per hour |
| 0.820208 | varas (Texas) per minute |
| 0.0136701 | varas (Texas) per second |
| 3,280.8 | feet per day |
| 136.7 | feet per hour |
| 2.278333 | feet per minute |
| 0.0379722 | feet per second |
| 4,374.440070 | spans per day |
| 182.268336 | spans per hour |
| 3.0378056 | spans per minute |
| 0.0506301 | spans per second |
| 4,970.974310 | links per day |
| 207.123916 | links per hour |
| 3.452065 | links per minute |
| 0.0575344 | links per second |
| 9,842.50 | hands per day |
| 410.104166 | hands per hour |
| 6.835069 | hands per minute |
| 0.113918 | hands per second |
| 10,000 | decimeters per day |
| 416.666664 | decimeters per hour |
| 6.944444 | decimeters per minute |
| 0.115741 | decimeters per second |
| 100,000 | centimeters per day |
| 4,166.666640 | centimeters per hour |
| 69.444444 | centimeters per minute |
| 1.157407 | centimeters per second |

## KILOMETER PER DAY (cont'd): =

| | |
|---|---|
| 39,370 | inches per day |
| 1,640.416666 | inches per hour |
| 27.340278 | inches per minute |
| 0.455671 | inches per second |
| 1,000,000 | millimeters per day |
| 41,666.6664 | millimeters per hour |
| 694.444444 | millimeters per minute |
| 11.574074 | millimeters per second |

## KILOMETER PER HOUR: =

| | |
|---|---|
| 14.912880 | miles per day |
| 0.62137 | miles per hour |
| 0.0103562 | miles per minute |
| 0.000172603 | miles per second |
| 24 | kilometers per day |
| 1 | kilometers per hour |
| 0.0166667 | kilometers per minute |
| 0.000277778 | kilometers per second |
| 119.303376 | furlongs per day |
| 4.970974 | furlongs per hour |
| 0.0828496 | furlongs per minute |
| 0.00138083 | furlongs per second |
| 240 | hektometers per day |
| 10 | hektometers per hour |
| 0.166667 | hektometers per minute |
| 0.00277778 | hektometers per second |
| 1,193.0337599 | chains per day |
| 49.709741 | chains per hour |
| 0.828496 | chains per minute |
| 0.0138083 | chains per second |
| 2,400 | dekameters per day |
| 100 | dekameters per hour |
| 1.666667 | dekameters per minute |
| 0.0277778 | dekameters per second |
| 4,772.125895 | rods per day |
| 198.838579 | rods per hour |
| 3.313976 | rods per minute |
| 0.0552329 | rods per second |
| 24,000 | meters per day |
| 1,000 | meters per hour |
| 16.666667 | meters per minute |
| 0.277778 | meters per second |

KILOMETER PER HOUR (cont'd): =

|  |  |
|---|---|
| 26,246.4 | yards per day |
| 1,093.6 | yards per hour |
| 18.226667 | yards per minute |
| 0.303777 | yards per second |
| 28,346.399997 | varas (Texas) per day |
| 1,181.1 | varas (Texas) per hour |
| 19.685 | varas (Texas) per minute |
| 0.328083 | varas (Texas) per second |
| 78,739.199997 | feet per day |
| 3,280.8 | feet per hour |
| 54.68 | feet per minute |
| 0.911333 | feet per second |
| 104,986.561622 | spans per day |
| 4,374.440070 | spans per hour |
| 72.907334 | spans per minute |
| 1.215122 | spans per second |
| 119,303.375990 | links per day |
| 4,970.974310 | links per hour |
| 82.849567 | links per minute |
| 1.380826 | links per second |
| 236,219.999933 | hands per day |
| 9,842.5 | hands per hour |
| 164.0416666 | hands per minute |
| 2.734028 | hands per second |
| 240,000 | decimeters per day |
| 10,000 | decimeters per hour |
| 166.666667 | decimeters per minute |
| 2.777778 | decimeters per second |
| 2,400,000 | centimeters per day |
| 100,000 | centimeters per hour |
| 1,666.666667 | centimeters per minute |
| 27.777778 | centimeters per second |
| 944,879.999904 | inches per day |
| 39,370 | inches per hour |
| 656.166667 | inches per minute |
| 10.936111 | inches per second |
| 24,000,000 | millimeters per day |
| 1,000,000 | millimeters per hour |
| 16,666.666667 | millimeters per minute |
| 277.777778 | millimeters per second |
| 0.5396 | knots |

KILOMETER PER MINUTE: =

| | |
|---|---|
| 894.772800 | miles per day |
| 37.282200 | miles per hour |
| 0.62137 | miles per minute |
| 0.0103562 | miles per second |
| 1,440 | kilometers per day |
| 60 | kilometers per hour |
| 1 | kilometers per minute |
| 0.0166667 | kilometers per second |
| 7,158.202560 | furlongs per day |
| 298.258440 | furlongs per hour |
| 4.970974 | furlongs per minute |
| 0.0828496 | furlongs per second |
| 14,400 | hektometers per day |
| 600 | hektometers per hour |
| 10 | hektometers per minute |
| 0.166667 | hektometers per second |
| 71,582.0256029 | chains per day |
| 2,982.584400 | chains per hour |
| 49.709741 | chains per minute |
| 0.828496 | chains per second |
| 144,000 | dekameters per day |
| 6,000 | dekameters per hour |
| 100 | dekameters per minute |
| 1.666667 | dekameters per second |
| 286,327.553184 | rods per day |
| 11,930.314716 | rods per hour |
| 198.838579 | rods per minute |
| 3.313976 | rods per second |
| 1,440,000 | meters per day |
| 60,000 | meters per hour |
| 1,000 | meters per minute |
| 16.666667 | meters per second |
| 1,574,784 | yards per day |
| 65,616.000012 | yards per hour |
| 1,093.6 | yards per minute |
| 18.226667 | yards per second |
| 1,700,784 | varas (Texas) per day |
| 70,866.0 | varas (Texas) per hour |
| 1,181.1 | varas (Texas) per minute |
| 19.685 | varas (Texas) per second |
| 4,724,352 | feet per day |
| 196,848 | feet per hour |
| 3,280.8 | feet per minute |
| 54.68 | feet per second |

**KILOMETER PER MINUTE (cont'd): =**

| | |
|---|---|
| 6,299,194 | spans per day |
| 262,466.404200 | spans per hour |
| 4,374.440070 | spans per minute |
| 72.907335 | spans per second |
| 7,158,203 | links per day |
| 298,258.440012 | links per hour |
| 4,970.974310 | links per minute |
| 82.849567 | links per second |
| 14,173,200 | hands per day |
| 590,550 | hands per hour |
| 9,842.5 | hands per minute |
| 164.0416667 | hands per second |
| 14,400,000 | decimeters per day |
| 600,000 | decimeters per hour |
| 10,000 | decimeters per minute |
| 166.666667 | decimeters per second |
| 144,000,000 | centimeters per day |
| 6,000,000 | centimeters per hour |
| 100,000 | centimeters per minute |
| 1,666.666667 | centimeters per second |
| 56,692,800 | inches per day |
| 2,362,200 | inches per hour |
| 39,370 | inches per minute |
| 656.166667 | inches per second |
| 1,440,000,000 | millimeters per day |
| 60,000,000 | millimeters per hour |
| 1,000,000 | millimeters per minute |
| 16,666.666667 | millimeters per second |

**KILOMETER PER SECOND: =**

| | |
|---|---|
| 53,686.3680 | miles per day |
| 2,236.93200 | miles per hour |
| 37.28220 | miles per minute |
| 0.62137 | miles per second |
| 86,400 | kilometers per day |
| 3,600 | kilometers per hour |
| 60 | kilometers per minute |
| 1 | kilometers per second |
| 429,492.180384 | furlongs per day |
| 17,895.506400 | furlongs per hour |
| 298.258440 | furlongs per minute |
| 4.970974 | furlongs per second |

## KILOMETER PER SECOND (cont'd): =

| | |
|---|---|
| 864,000 | hektometers per day |
| 36,000 | hektometers per hour |
| 600 | hektometers per minute |
| 10 | hektometers per second |
| 4,294,922 | chains per day |
| 178,955.064 | chains per hour |
| 2,982.58440 | chains per minute |
| 49.709741 | chains per second |
| 8,640,000 | dekameters per day |
| 360,000 | dekameters per hour |
| 6,000 | dekameters per minute |
| 100 | dekameters per second |
| 17,179,653 | rods per day |
| 715,818.88440 | rods per hour |
| 11,930.314740 | rods per minute |
| 198.838579 | rods per second |
| 86,400,000 | meters per day |
| 3,600,000 | meters per hour |
| 60,000 | meters per minute |
| 1,000 | meters per second |
| 94,487,040 | yards per day |
| 3,936,960 | yards per hour |
| 65,616 | yards per minute |
| 1,093.6 | yards per second |
| 102,047,040 | varas (Texas) per day |
| 4,251,960 | varas (Texas) per hour |
| 70,866 | varas (Texas) per minute |
| 1,181.1 | varas (Texas) per second |
| 283,461,120 | feet per day |
| 11,810,880 | feet per hour |
| 196,848 | feet per minute |
| 3,280.8 | feet per second |
| 377,951,622 | spans per day |
| 15,747,984 | spans per hour |
| 262,466.404200 | spans per minute |
| 4,374.440070 | spans per second |
| 429,492,180 | links per day |
| 17,895,508 | links per hour |
| 298,258.440 | links per minute |
| 4,970.974310 | links per second |
| 850,392,000 | hands per day |
| 35,433,000 | hands per hour |
| 590,550 | hands per minute |
| 9,842.5 | hands per second |

KILOMETER PER SECOND (cont'd): =

| | |
|---|---|
| 864,000,000 | decimeters per day |
| 36,000,000 | decimeters per hour |
| 600,000 | decimeters per minute |
| 10,000 | decimeters per second |
| 8,640,000,000 | centimeters per day |
| 360,000,000 | centimeters per hour |
| 6,000,000 | centimeters per minute |
| 100,000 | centimeters per second |
| 3,401,568,000 | inches per day |
| 141,732,000 | inches per hour |
| 2,362,200 | inches per minute |
| 39,370 | inches per second |
| 86,400,000,000 | millimeters per day |
| 3,600,000,000 | millimeters per hour |
| 60,000,000 | millimeters per minute |
| 1,000,000 | millimeters per second |

KILOMETER PER HOUR PER SECOND: =

| | |
|---|---|
| 0.000172594 | miles per second per second |
| 0.0002778 | kilometers per second per second |
| 0.002778 | hektometers per second per second |
| 0.02778 | dekameters per second per second |
| 0.2778 | meters per second per second |
| 0.303767 | yards per second per second |
| 0.9113 | feet per second per second |
| 2.778 | decimeters per second per second |
| 27.78 | centimeters per second per second |
| 10.9356 | inches per second per second |
| 277.8 | millimeters per second per second |

KILOLITER: =

| | |
|---|---|
| 1 | kiloliters |
| 1 | cubic meters |
| 1.3080 | cubic yards |
| 10 | hektoliters |
| 28.378 | bushels (U.S.) dry |
| 27.497 | bushels (Imperial) dry |
| 35.316 | cubic feet |
| 100 | dekaliters |
| 11.3513 | pecks |

KILOLITER (cont'd): =

| | |
|---|---|
| 264.18 | gallons (U.S.) liquid |
| 227.0574264 | gallons (U.S.) dry |
| 219.977402 | gallons (Imperial) |
| 1,056.72 | quarts (liquid) |
| 908.110825 | quarts (dry) |
| 1,000 | liters |
| 1,000 | cubic decimeters |
| 2,113.44 | pints (liquid) |
| 8,453.76 | gills (liquid) |
| 10,000 | deciliters |
| 34,607.58 | cubic inches |
| 100,000 | centiliters |
| 1,000,000 | milliliters |
| 1,000,000 | cubic centimeters |
| 1,000,000,000 | cubic millimeters |
| 1.101234 | tons (short) of water @ 62° F. |
| 0.983252 | tons (long) of water @ 62° F. |
| 1.0 | tons (metric) of water @ 62° F. |
| 1,000 | kilograms of water @ 62° F. |
| 2,202.46866 | pounds (Avoir.) of water @ 62° F. |
| 2,676.611 | pounds (Troy) of water @ 62° F. |
| 10,000 | hektograms of water @ 62° F. |
| 100,000 | dekagrams of water @ 62° F. |
| 32,119.331983 | ounces (Troy) of water @ 62° F. |
| 35,239.49856 | ounces (Avoir.) of water @ 62° F. |
| 256,954.655866 | drams (Troy) of water @ 62° F. |
| 563,831.97696 | drams (Avoir.) of water @ 62° F. |
| 642,386.639664 | pennyweights of water @ 62° F. |
| 770,863.967597 | scruples (Avoir.) |
| 1,000,000 | grams of water @ 62° F. |
| 10,000,000 | decigrams of water @ 62° F. |
| 15,417,281 | grains of water @ 62° F. |
| 100,000,000 | centigrams of water @ 62° F. |
| 1,000,000,000 | milligrams of water @ 62° F. |
| 33,815.04 | ounces (fluid) |
| 270,520.32 | drams (fluid) |
| 32.105795 | sacks of cement |

KILOWATT: =

| | |
|---|---|
| 63,725,184 | foot pounds per day |
| 2,655,216 | foot pounds per hour |
| 44,253.60 | foot pounds per minute |

KILOWATT (cont'd): =

| | |
|---|---|
| 737.56 | foot pounds per second |
| 764,702,208 | inch pounds per day |
| 31,862,592 | inch pounds per hour |
| 531,043.20 | inch pounds per minute |
| 8,850.72 | inch pounds per second |
| 20,640.09600 | kilogram calories (mean) per day |
| 860.004 | kilogram calories (mean) per hour |
| 14.33340 | kilogram calories (mean) per minute |
| 0.23889 | kilogram calories (mean) per second |
| 45,503.615916 | pound calories (mean) per day |
| 1,895.983996 | pound calories (mean) per hour |
| 31.599733 | pound calories (mean) per minute |
| 0.526662 | pound calories (mean) per second |
| 728,057.85472 | ounce calories (mean) per day |
| 30,335.743936 | ounce calories (mean) per hour |
| 505.595728 | ounce calories (mean) per minute |
| 8.426592 | ounce calories (mean) per second |
| 20,640,096 | gram calories (mean) per day |
| 860,004 | gram calories (mean) per hour |
| 14,333.40 | gram calories (mean) per minute |
| 238.89 | gram calories (mean) per second |
| 81,930.52800 | BTU (mean) per day |
| 3,413.77200 | BTU (mean) per hour |
| 56.89620 | BTU (mean) per minute |
| 0.94827 | BTU (mean) per second |
| 1 | kilowatts |
| 1,000 | watts |
| 3,600,000 | joules |
| 1.341 | horsepower |
| 1.3597 | horsepower (metric) |
| 1.3597 | Cheval-Vapeur hours |
| 0.234 | pounds carbon oxidized with 100% efficiency |
| 3.52 | pounds water evaporated from and at 212° F. |
| 852.647040 | kiloliter-atmospheres per day |
| 35.52695 | kiloliter-atmospheres per hour |
| 0.592116 | kiloliter-atmospheres per minute |
| 0.0098686 | kiloliter-atmospheres per second |
| 852,647 | liter-atmospheres per day |
| 35,526.95 | liter-atmospheres per hour |
| 592.116 | liter-atmospheres per minute |
| 9,8686 | liter-atmospheres per second |
| 8,808,000 | kilogram meters per day |
| 367,000 | kilogram meters per hour |
| 6,116.666667 | kilogram meters per minute |
| 101.944444 | kilogram meters per second |

LINK (SURVEYORS): =

| | |
|---|---|
| 0.0001250 | miles |
| 0.000201168 | kilometers |
| 0.001 | furlongs |
| 0.00201168 | hektometers |
| 0.01 | chains |
| 0.0201168 | dekameters |
| 0.04 | rods |
| 0.201168 | meters |
| 0.22 | yards |
| 0.23760 | varas (Texas) |
| 0.66 | feet |
| 0.879998 | spans |
| 1.98 | hands |
| 2.011684 | decimeters |
| 20.11684 | centimeters |
| 7.92 | inches |
| 201.1684 | millimeters |
| 7,920 | mils |
| 201,168 | microns |
| 201,168,400 | milli-microns |
| 201,168,400 | micro-millimeters |
| 312,447 | wave lengths of red line of cadmium |
| 201,168,400,000 | Angstrom Units |

LITER: =

| | |
|---|---|
| 0.001 | kiloliters |
| 0.001 | cubic meters |
| 0.0013080 | cubic yards |
| 0.00628995 | barrels |
| 0.01 | hektoliters |
| 0.028378 | bushels (U.S.) dry |
| 0.027497 | bushels (Imperial) dry |
| 0.0353144 | cubic feet |
| 0.1 | dekaliters |
| 0.113512 | pecks (U.S.) dry |
| 0.264178 | gallons (U.S.) liquid |
| 0.22702 | gallons (U.S.) dry |
| 0.21998 | gallons (Imperial) |
| 1.056710 | quarts (liquid) |
| 0.908102 | quarts (dry) |
| 1 | cubic decimeters |
| 1.8162 | pints (U.S.) dry |
| 2.1134 | pints (U.S.) liquid |

LITER (cont'd): =

| | |
|---|---|
| 7.0392 | gills (Imperial) |
| 8.4538 | gills (U.S.) |
| 10 | deciliters |
| 61.025 | cubic inches |
| 100 | centiliters |
| 1,000 | milliliters |
| 1,000 | cubic centimeters |
| 1,000,000 | cubic millimeters |
| 33.8147 | ounces (U.S.) fluid |
| 35.196 | ounces (Imperial) fluid |
| 270.5179 | drams (fluid) |
| 16,231.0740 | minims |
| 2.20462 | pounds of water at maximum density |

LITER ATMOSPHERE: =

| | |
|---|---|
| 2,404.59243 | foot poundals |
| 28,855.10916 | inch poundals |
| 74.738589 | foot pounds |
| 896.863068 | inch pounds |
| 0.0000242071 | ton calories |
| 0.0242071 | kilogram calories |
| 0.0533676 | pound calories |
| 0.242071 | hektogram calories |
| 2.42071 | dekagram calories |
| 0.853881 | ounce calories |
| 24.2071 | gram calories (mean) |
| 242.071 | decigram calories |
| 2,420.71 | centigram calories |
| 24,207.1 | milligram calories |
| 0.00000117258 | kilowatt days |
| 0.0000281419 | kilowatt hours |
| 0.00168852 | kilowatt minutes |
| 0.101311 | kilowatt seconds |
| 0.00117258 | watt days |
| 0.0281419 | watt hours |
| 1.688516 | watt minutes |
| 101.310932 | watt seconds |
| 0.010333 | ton meters |
| 10.333 | kilogram meters |
| 22.780362 | pound meters |
| 103.33 | hektogram meters |
| 1,033.3 | dekagram meters |

| | |
|---|---|
| 364.485798 | ounce meters |
| 10,333 | gram meters |
| 103,330 | decigram meters |
| 1,033,300 | centigram meters |
| 10,333,000 | milligram meters |
| 0.00010333 | ton hektometers |
| 0.10333 | kilogram hektometers |
| 0.227804 | pound hektometers |
| 1.0333 | hektogram hektometers |
| 10.333 | dekagram hektometers |
| 3.644858 | ounce hektometers |
| 103.33 | gram hektometers |
| 1,033.3 | decigram hektometers |
| 10,333 | centigram hektometers |
| 103,330 | milligram hektometers |
| 0.0010330 | ton dekameters |
| 1.0333 | kilogram dekameters |
| 2.278036 | pound dekameters |
| 10.333 | hektogram dekameters |
| 103.33 | dekagram dekameters |
| 36.448583 | ounce dekameters |
| 1033.3 | gram dekameters |
| 10,333 | decigram dekameters |
| 103,330 | centigram dekameters |
| 1,033,300 | milligram dekameters |
| 0.0339001 | ton feet |
| 33.900951 | kilogram feet |
| 74.738796 | pound feet |
| 339.009507 | hektogram feet |
| 3,390.095072 | dekagram feet |
| 1,195.820731 | ounce feet |
| 33,900.950719 | gram feet |
| 339,009.50719 | decigram feet |
| 3,390,095 | centigram feet |
| 33,900,951 | milligram feet |
| 0.406811 | ton inches |
| 406.811409 | kilogram inches |
| 896.865507 | pound inches |
| 4,068.114086 | hektogram inches |
| 40,681.140863 | dekagram inches |
| 14,349.848105 | ounce inches |
| 406,811.408628 | gram inches |
| 4,068,114 | decigram inches |
| 40,681,141 | centigram inches |
| 406,811,409 | milligram inches |

LITER ATMOSPHERE (cont'd): =

| | |
|---|---|
| 0.10333 | ton decimeters |
| 103.33 | kilogram decimeters |
| 227.803622 | pound decimeters |
| 1,033.3 | hektogram decimeters |
| 10,333 | dekagram decimeters |
| 3,644.857956 | ounce decimeters |
| 103,330 | gram decimeters |
| 1,033,300 | decigram decimeters |
| 10,333,000 | centigram decimeters |
| 103,330,000 | milligram decimeters |
| 1.0333 | ton centimeters |
| 1,033.3 | kilogram centimeters |
| 2,278.036223 | pound centimeters |
| 10,333 | hektogram centimeters |
| 103,330 | dekagram centimeters |
| 36,448.579561 | ounce centimeters |
| 1,033,300 | gram centimeters |
| 10,333,000 | decigram centimeters |
| 103,330,000 | centigram centimeters |
| 1,033,300,000 | milligram centimeters |
| 10.333 | ton millimeters |
| 10,333 | kilogram millimeters |
| 22,780.362226 | pound millimeters |
| 103,330 | hektogram millimeters |
| 1,033,300 | dekagram millimeters |
| 364,485.795614 | ounce millimeters |
| 10,333,000 | gram millimeters |
| 103,330,000 | decigram millimeters |
| 1,033,300,000 | centigram millimeters |
| 10,333,000,000 | milligram millimeters |
| 0.001 | kiloliter-atmospheres |
| 0.01 | hektoliter-atmospheres |
| 0.1 | dekaliter-atmospheres |
| 1 | liter-atmospheres |
| 10 | deciliter-atmospheres |
| 100 | centiliter-atmospheres |
| 1,000 | millimeter-atmospheres |
| 0.000000000001 | cubic kilometer-atmospheres |
| 0.000000001 | cubic hektometer-atmospheres |
| 0.000001 | cubic dekameter-atmospheres |
| 0.001 | cubic meter-atmospheres |
| 0.035319 | cubic foot-atmospheres |
| 61.025 | cubic inch-atmospheres |
| 1 | cubic decimeter-atmospheres |

**LITER ATMOSPHERE (cont'd): =**

| | |
|---|---|
| 1,000 | cubic centimeter-atmospheres |
| 1,000,000 | cubic millimeter-atmospheres |
| 101.328 | joules (absolute) |
| 0.00003827 | Cheval-Vapeur hours |
| 0.00000157277 | horsepower days |
| 0.000037745 | horsepower hours |
| 0.00226479 | horsepower minutes |
| 0.135887 | horsepower seconds |
| 0.00000658398 | pounds of carbon oxidized with 100% efficiency |
| 0.00009907 | pounds of water evaporated from and at 212° F. |
| 43.573724 | BTU (mean) per pound |
| 0.09607 | BTU (mean) |
| 1,013,321,145 | ergs |

**LITER PER DAY: =**

| | |
|---|---|
| 0.001 | kiloliters per day |
| 0.0000416667 | kiloliters per hour |
| 0.000000694444 | kiloliters per minute |
| 0.0000000115741 | kiloliters per second |
| 0.001 | cubic meters per day |
| 0.0000416667 | cubic meters per hour |
| 0.000000694444 | cubic meters per minute |
| 0.0000000115741 | cubic meters per second |
| 0.0013080 | cubic yards per day |
| 0.0000545000 | cubic yards per hour |
| 0.000000908333 | cubic yards per minute |
| 0.0000000151389 | cubic yards per second |
| 0.00628996 | barrels per day |
| 0.000262082 | barrels per hour |
| 0.00000436803 | barrels per minute |
| 0.0000000728005 | barrels per second |
| 0.01 | hektoliters per day |
| 0.000416667 | hektoliters per hour |
| 0.00000694444 | hektoliters per minute |
| 0.000000115741 | hektoliters per second |
| 0.028378 | bushels (U.S.—dry) per day |
| 0.00118242 | bushels (U.S.—dry) per hour |
| 0.0000197069 | bushels (U.S.—dry) per minute |
| 0.000000328449 | bushels (U.S.—dry) per second |
| 0.027497 | bushels (Imperial-dry) per day |
| 0.00114571 | bushels (Imperial-dry) per hour |
| 0.0000190951 | bushels (Imperial-dry) per minute |

LITER PER DAY (cont'd): =

0.000000318252 . . . . . . . . . . . . . . . . . bushels (Imperial-dry) per second
0.0353144 . . . . . . . . . . . . . . . . . . . . . . cubic feet per day
0.00147143 . . . . . . . . . . . . . . . . . . . . cubic feet per hour
0.000024539 . . . . . . . . . . . . . . . . . . . cubic feet per minute
0.000000408731 . . . . . . . . . . . . . . . . . cubic feet per second
0.1 . . . . . . . . . . . . . . . . . . . . . . . . . . dekaliters per day
0.00416667 . . . . . . . . . . . . . . . . . . . . .dekaliters per hour
0.0000694444 . . . . . . . . . . . . . . . . . . dekaliters per minute
0.00000115741 . . . . . . . . . . . . . . . . . dekaliters per second
0.113512 . . . . . . . . . . . . . . . . . . . .pecks (U.S.—dry) per day
0.00472967 . . . . . . . . . . . . . . . . . . pecks (U.S.—dry) per hour
0.0000788278 . . . . . . . . . . . . . . . .pecks (U.S.—dry) per minute
0.00000131380 . . . . . . . . . . . . . . . pecks (U.S.—dry) per second
0.264178 . . . . . . . . . . . . . . . . . . .gallons (U.S.—liquid) per day
0.0110074 . . . . . . . . . . . . . . . . . gallons (U.S.—liquid) per hour
0.000183457 . . . . . . . . . . . . . . . .gallons (U.S.—liquid) per minute
0.00000305762 . . . . . . . . . . . . . .gallons (U.S.—liquid) per second
0.22702 . . . . . . . . . . . . . . . . . . . . . . gallons (U.S.—dry) per day
0.00945917 . . . . . . . . . . . . . . . . . . gallons (U.S.—dry) per hour
0.000157653 . . . . . . . . . . . . . . . . . gallons (U.S.—dry) per minute
0.00000262755 . . . . . . . . . . . . . . . gallons (U.S.—dry) per second
0.21998 . . . . . . . . . . . . . . . . . . . . . .gallons (Imperial) per day
0.00916583 . . . . . . . . . . . . . . . . . . gallons (Imperial) per hour
0.000152764 . . . . . . . . . . . . . . . . .gallons (Imperial) per minute
0.00000254606 . . . . . . . . . . . . . . . .gallons (Imperial) per second
1.056710 . . . . . . . . . . . . . . . . . . . . . quarts (liquid) per day
0.0440296 . . . . . . . . . . . . . . . . . . . .quarts (liquid) per hour
0.000733826 . . . . . . . . . . . . . . . . . quarts (liquid) per minute
0.0000122304 . . . . . . . . . . . . . . . . .quarts (liquid) per second
0.908102 . . . . . . . . . . . . . . . . . . . . . . quarts (dry) per day
0.0378376 . . . . . . . . . . . . . . . . . . . . . quarts (dry) per hour
0.000630626 . . . . . . . . . . . . . . . . . . .quarts (dry) per minute
0.0000105104 . . . . . . . . . . . . . . . . . . quarts (dry) per second
1 . . . . . . . . . . . . . . . . . . . . . . . . . . . . . liters per day
0.0416667 . . . . . . . . . . . . . . . . . . . . . . . liters per hour
0.000694444 . . . . . . . . . . . . . . . . . . . . . liters per minute
0.0000115741 . . . . . . . . . . . . . . . . . . . . liters per second
1 . . . . . . . . . . . . . . . . . . . . . . . cubic decimeters per day
0.0416667 . . . . . . . . . . . . . . . . . cubic decimeters per hour
0.000694444 . . . . . . . . . . . . . . . cubic decimeters per minute
0.0000115741 . . . . . . . . . . . . . . cubic decimeters per second
1.8162 . . . . . . . . . . . . . . . . . . . . . . pints (U.S.—dry) per day
0.0756750 . . . . . . . . . . . . . . . . . . . pints (U.S.—dry) per hour
0.00126125 . . . . . . . . . . . . . . . . . pints (U.S.—dry) per minute

LITER PER DAY (cont'd): =

0.0000210208 . . . . . . . . . . . . . . . . . . . . pints (U.S.—dry) per second
2.1134 . . . . . . . . . . . . . . . . . . . . . . . pints (U.S.—liquid) per day
0.0880583 . . . . . . . . . . . . . . . . . . . . . pints (U.S.—liquid) per hour
0.00146764 . . . . . . . . . . . . . . . . . . . . pints (U.S.—liquid) per minute
0.0000244606 . . . . . . . . . . . . . . . . . . . pints (U.S.—liquid) per second
7.0392 . . . . . . . . . . . . . . . . . . . . . . . gills (Imperial) per day
0.293300 . . . . . . . . . . . . . . . . . . . . . gills (Imperial) per hour
0.00488833 . . . . . . . . . . . . . . . . . . . . gills (Imperial) per minute
0.0000814722 . . . . . . . . . . . . . . . . . . . gills (Imperial) per second
8.4538 . . . . . . . . . . . . . . . . . . . . . . . gills (U.S.) per day
0.352242 . . . . . . . . . . . . . . . . . . . . . gills (U.S.) per hour
0.00587069 . . . . . . . . . . . . . . . . . . . . gills (U.S.) per minute
0.0000978449 . . . . . . . . . . . . . . . . . . . gills (U.S.) per second
10 . . . . . . . . . . . . . . . . . . . . . . . . . deciliters per day
0.416667 . . . . . . . . . . . . . . . . . . . . . deciliters per hour
0.00694444 . . . . . . . . . . . . . . . . . . . . deciliters per minute
0.000115741 . . . . . . . . . . . . . . . . . . . deciliters per second
61.025 . . . . . . . . . . . . . . . . . . . . . . . cubic inches per day
2.542708 . . . . . . . . . . . . . . . . . . . . . cubic inches per hour
0.0423785 . . . . . . . . . . . . . . . . . . . . . cubic inches per minute
0.000706308 . . . . . . . . . . . . . . . . . . . cubic inches per second
100 . . . . . . . . . . . . . . . . . . . . . . . . centiliters per day
4.166667 . . . . . . . . . . . . . . . . . . . . . centiliters per hour
0.0694444 . . . . . . . . . . . . . . . . . . . . . centiliters per minute
0.00115741 . . . . . . . . . . . . . . . . . . . . centiliters per second
1,000 . . . . . . . . . . . . . . . . . . . . . . . milliliters per day
41.666667 . . . . . . . . . . . . . . . . . . . . . milliliters per hour
0.694444 . . . . . . . . . . . . . . . . . . . . . milliliters per minute
0.0115741 . . . . . . . . . . . . . . . . . . . . . milliliters per second
1,000 . . . . . . . . . . . . . . . . . . . . . . . cubic centimeters per day
41.666667 . . . . . . . . . . . . . . . . . . . . . cubic centimeters per hour
0.694444 . . . . . . . . . . . . . . . . . . . . . cubic centimeters per minute
0.0115741 . . . . . . . . . . . . . . . . . . . . . cubic centimeters per second
1,000,000 . . . . . . . . . . . . . . . . . . . . . cubic millimeters per day
41,666.666400 . . . . . . . . . . . . . . . . . . cubic millimeters per hour
694.444444 . . . . . . . . . . . . . . . . . . . . cubic millimeters per minute
11.574074 . . . . . . . . . . . . . . . . . . . . . cubic millimeters per second
33.8147 . . . . . . . . . . . . . . . . . . . . . . ounces (U.S.) fluid per day
1.408946 . . . . . . . . . . . . . . . . . . . . . ounces (U.S.) fluid per hour
0.0234824 . . . . . . . . . . . . . . . . . . . . . ounces (U.S.) fluid per minute
0.000391374 . . . . . . . . . . . . . . . . . . . ounces (U.S.) fluid per second
35.196 . . . . . . . . . . . . . . . . . . . . . . . ounces (Imperial—fluid) per day
1.466500 . . . . . . . . . . . . . . . . . . . . . ounces (Imperial—fluid) per hour
0.0244417 . . . . . . . . . . . . . . . . . . . . . ounces (Imperial—fluid) per minute

**LITER PER DAY (cont'd): =**

| | |
|---|---|
| 0.000407361 | ounces (Imperial–fluid) per second |
| 270.5179 | drams (fluid) per day |
| 11.271579 | drams (fluid) per hour |
| 0.187860 | drams (fluid) per minute |
| 0.00313099 | drams (fluid) per second |
| 16,231.0740 | minims per day |
| 676.294747 | minims per hour |
| 11.271579 | minims per minute |
| 0.187860 | minims per second |

**LITER PER HOUR: =**

| | |
|---|---|
| 0.0240 | kiloliters per day |
| 0.001 | kiloliters per hour |
| 0.0000166667 | kiloliters per minute |
| 0.000000277778 | kiloliters per second |
| 0.0240 | cubic meters per day |
| 0.001 | cubic meters per hour |
| 0.0000166667 | cubic meters per minute |
| 0.000000277778 | cubic meters per second |
| 0.031392 | cubic yards per day |
| 0.0013080 | cubic yards per hour |
| 0.0000218 | cubic yards per minute |
| 0.000000363333 | cubic yards per second |
| 0.150959 | barrels per day |
| 0.00628995 | barrels per hour |
| 0.000104833 | barrels per minute |
| 0.00000174721 | barrels per second |
| 0.24 | hektoliters per day |
| 0.01 | hektoliters per hour |
| 0.000166667 | hektoliters per minute |
| 0.00000277778 | hektoliters per second |
| 0.681072 | bushels (U.S.–dry) per day |
| 0.028378 | bushels (U.S.–dry) per hour |
| 0.000472967 | bushels (U.S.–dry) per minute |
| 0.00000788278 | bushels (U.S.–dry) per second |
| 0.659928 | bushels (Imperial–dry) per day |
| 0.027497 | bushels (Imperial–dry) per hour |
| 0.000458283 | bushels (Imperial–dry) per minute |
| 0.00000763806 | bushels (Imperial–dry) per second |
| 0.84746 | cubic feet per day |
| 0.0353144 | cubic feet per hour |

LITER PER HOUR (cont'd): =

| | |
|---|---|
| 0.000588573 | cubic feet per minute |
| 0.00000980956 | cubic feet per second |
| 2.4 | dekaliters per day |
| 0.1 | .dekaliters per hour |
| 0.00166667 | dekaliters per minute |
| 0.0000277778 | dekaliters per second |
| 2.724288 | .pecks (U.S.—dry) per day |
| 0.113512 | pecks (U.S.—dry) per hour |
| 0.00189187 | .pecks (U.S.—dry) per minute |
| 0.0000315311 | pecks (U.S.—dry) per second |
| 6.340272 | .gallons (U.S.—liquid) per day |
| 0.264178 | gallons (U.S.—liquid) per hour |
| 0.00440297 | .gallons (U.S.—liquid) per minute |
| 0.0000733828 | .gallons (U.S.—liquid) per second |
| 5.448480 | gallons (U.S.—dry) per day |
| 0.22702 | gallons (U.S.—dry) per hour |
| 0.00378367 | gallons (U.S.—dry) per minute |
| 0.0000630611 | gallons (U.S.—dry) per second |
| 5.279520 | .gallons (Imperial) per day |
| 0.21998 | gallons (Imperial) per hour |
| 0.00366633 | .gallons (Imperial) per minute |
| 0.0000611056 | .gallons (Imperial) per second |
| 25.361040 | quarts (liquid) per day |
| 1.056710 | quarts (liquid) per hour |
| 0.0176118 | quarts ((liquid) per minute |
| 0.000293531 | .quarts (liquid) per second |
| 21.794448 | quarts (dry) per day |
| 0.908102 | quarts (dry) per hour |
| 0.0151350 | quarts (dry) per minute |
| 0.000252251 | quarts (dry) per second |
| 24 | liters per day |
| 1 | liters per hour |
| 0.0166667 | liters per minute |
| 0.000277778 | liters per second |
| 24 | cubic decimeters per day |
| 1 | cubic decimeters per hour |
| 0.0166667 | cubic decimeters per minute |
| 0.00027778 | cubic decimeters per second |
| 43.58880 | pints (U.S.—dry) per day |
| 1.8162 | pints (U.S.—dry) per hour |
| 0.0302700 | pints (U.S.—dry) per minute |
| 0.0005045 | .pints (U.S.)dry) per second |
| 50.721600 | .pints (U.S.—liquid) per day |
| 2.1134 | pints (U.S.—liquid) per hour |

LITER PER HOUR (cont'd): =

| | |
|---|---|
| 0.0352233 | pints (U.S.—liquid) per minute |
| 0.000587056 | pints (U.S.—liquid) per second |
| 168.940800 | gills (Imperial) per day |
| 7.0392 | gills (Imperial) per hour |
| 0.117320 | gills (Imperial) per minute |
| 0.00195533 | gills (Imperial) per second |
| 202.891200 | gills (U.S.) per day |
| 8.4538 | gills (U.S.) per hour |
| 0.140897 | gills (U.S.) per minute |
| 0.00234828 | gills (U.S.) per second |
| 240 | deciliters per day |
| 10 | deciliters per hour |
| 0.166667 | deciliters per minute |
| 0.00277778 | deciliters per second |
| 1,464.6 | cubic inches per day |
| 61.025 | cubic inches per hour |
| 1.0170833 | cubic inches per minute |
| 0.0169514 | cubic inches per second |
| 2,400 | centiliters per day |
| 100 | centiliters per hour |
| 1.666667 | centiliters per minute |
| 0.0277778 | centiliters per second |
| 24,000 | milliliters per day |
| 1,000 | milliliters per hour |
| 16.666667 | milliliters per minute |
| 0.277778 | milliliters per second |
| 24,000 | cubic centimeters per day |
| 1,000 | cubic centimeters per hour |
| 16.666667 | cubic centimeters per minute |
| 0.277778 | cubic centimeters per second |
| 24,000,000 | cubic millimeters per day |
| 1,000,000 | cubic millimeters per hour |
| 16,666.666667 | cubic millimeters per minute |
| 277.777778 | cubic millimeters per second |
| 811.552800 | ounces (U.S.) fluid per day |
| 33.8147 | ounces (U.S.) fluid per hour |
| 0.563578 | ounces (U.S.) fluid per minute |
| 0.00939297 | ounces (U.S.) fluid per second |
| 844.704 | ounces (Imperial—fluid) per day |
| 35.196 | ounces (Imperial—fluid) per hour |
| 0.586600 | ounces (Imperial—fluid) per minute |
| 0.00977667 | ounces (Imperial—fluid) per second |
| 6,492.429599 | drams (fluid) per day |
| 270.5179 | drams (fluid) per hour |

LITER PER HOUR (cont'd): =

4.508632 . . . . . . . . . . . . . . . . . . . . . . . . . drams (fluid) per minute
0.0751439 . . . . . . . . . . . . . . . . . . . . . . . . drams (fluid) per second
389,545.775424 . . . . . . . . . . . . . . . . . . . . . .minims per day
16,231.0740 . . . . . . . . . . . . . . . . . . . . . . . minims per hour
270.5179 . . . . . . . . . . . . . . . . . . . . . . . . .minims per minute
4.508632 . . . . . . . . . . . . . . . . . . . . . . . . .minims per second

LITER PER MINUTE: =

1.440 . . . . . . . . . . . . . . . . . . . . . . . . . . . .kiloliters per day
0.0600 . . . . . . . . . . . . . . . . . . . . . . . . . . . kiloliters per hour
0.001 . . . . . . . . . . . . . . . . . . . . . . . . . . . .kiloliters per minute
0.0000166667 . . . . . . . . . . . . . . . . . . . . . . . kiloliters per second
1.440 . . . . . . . . . . . . . . . . . . . . . . . . . . . cubic meters per day
0.0600 . . . . . . . . . . . . . . . . . . . . . . . . . . cubic meters per hour
0.001 . . . . . . . . . . . . . . . . . . . . . . . . . . cubic meters per minute
0.0000166667 . . . . . . . . . . . . . . . . . . . . . .cubic meters per second
1.883520 . . . . . . . . . . . . . . . . . . . . . . . . . cubic yards per day
0.0784800 . . . . . . . . . . . . . . . . . . . . . . . . .cubic yards per hour
0.0013080 . . . . . . . . . . . . . . . . . . . . . . . cubic yards per minute
0.0000218 . . . . . . . . . . . . . . . . . . . . . . . . cubic yards per second
9.0575383 . . . . . . . . . . . . . . . . . . . . . . . . . . barrels per day
0.377397 . . . . . . . . . . . . . . . . . . . . . . . . . . barrels per hour
0.00628996 . . . . . . . . . . . . . . . . . . . . . . . . barrels per minute
0.000104833 . . . . . . . . . . . . . . . . . . . . . . . barrels per second
14.40 . . . . . . . . . . . . . . . . . . . . . . . . . . . .hektoliters per day
0.600 . . . . . . . . . . . . . . . . . . . . . . . . . . . hektoliters per hour
0.01 . . . . . . . . . . . . . . . . . . . . . . . . . . . .hektoliters per minute
0.000166667 . . . . . . . . . . . . . . . . . . . . . . .hektoliters per second
40.864320 . . . . . . . . . . . . . . . . . . . . . .bushels (U.S.—dry) per day
1.702680 . . . . . . . . . . . . . . . . . . . . . . bushels (U.S.—dry) per hour
0.028378 . . . . . . . . . . . . . . . . . . . . . .bushels (U.S.—dry) per minute
0.000472967 . . . . . . . . . . . . . . . . . . . . bushels (U.S.—dry) per second
39.595680 . . . . . . . . . . . . . . . . . . . bushels (Imperial—dry) per day
1.649820 . . . . . . . . . . . . . . . . . . . . .bushels (Imperial—dry) per hour
0.027497 . . . . . . . . . . . . . . . . . . . bushels (Imperial—dry) per minute
0.000458283 . . . . . . . . . . . . . . . . . bushels (Imperial—dry) per second
50.852736 . . . . . . . . . . . . . . . . . . . . . . . . . cubic feet per day
2.118864 . . . . . . . . . . . . . . . . . . . . . . . . . cubic feet per hour
0.0353144 . . . . . . . . . . . . . . . . . . . . . . . . cubic feet per minute
0.000588573 . . . . . . . . . . . . . . . . . . . . . . . cubic feet per second
144.0 . . . . . . . . . . . . . . . . . . . . . . . . . . . dekaliters per day
6.0 . . . . . . . . . . . . . . . . . . . . . . . . . . . . .dekaliters per hour

LITER PER MINUTE (cont'd): =

| | |
|---|---|
| 0.1 | dekaliters per minute |
| 0.00166667 | dekaliters per second |
| 163.457280 | pecks (U.S.—dry) per day |
| 6.810720 | pecks (U.S.—dry) per hour |
| 0.113512 | pecks (U.S.—dry) per minute |
| 0.00189187 | pecks (U.S.—dry) per second |
| 380.416320 | gallons (U.S.—liquid) per day |
| 15.850680 | gallons (U.S.—liquid) per hour |
| 0.264178 | gallons (U.S.—liquid) per minute |
| 0.00440297 | gallons (U.S.—liquid) per second |
| 326.908800 | gallons (U.S.—dry) per day |
| 13.621200 | gallons (U.S.—dry) per hour |
| 0.22702 | gallons (U.S.—dry) per minute |
| 0.00378367 | gallons (U.S.—dry) per second |
| 316.771200 | gallons (Imperial) per day |
| 13.198800 | gallons (Imperial) per hour |
| 0.21998 | gallons (Imperial) per minute |
| 0.00366633 | gallons (Imperial) per second |
| 1,521.662400 | quarts (liquid) per day |
| 63.402600 | quarts (liquid) per hour |
| 1.056710 | quarts (liquid) per minute |
| 0.0176118 | qurts (liquid) per second |
| 1,307.666880 | quarts (dry) per day |
| 54.486120 | quarts (dry) per hour |
| 0.908102 | quarts (dry) per minute |
| 0.0151350 | quarts (dry) per second |
| 1,440 | liters per day |
| 60 | liters per hour |
| 1 | liters per minute |
| 0.0166667 | liters per second |
| 1,440 | cubic decimeters per day |
| 60 | cubic decimeters per hour |
| 1 | cubic decimeters per minute |
| 0.0166667 | cubic decimeters per second |
| 2,615.328 | pints (U.S.—dry) per day |
| 108.972 | pints (U.S.—dry) per hour |
| 1.8162 | pints (U.S.—dry) per minute |
| 0.0302700 | pints (U.S.—dry) per second |
| 3,043.296 | pints (U.S.—liquid) per day |
| 126.804 | pints (U.S.—liquid) per hour |
| 2.1134 | pints (U.S.—liquid) per minute |
| 0.0352233 | pints (U.S.—liquid) per second |
| 10,136.448 | gills (Imperial) per day |
| 422.352 | gills (Imperial) per hour |

LITER PER MINUTE (cont'd): =

| | |
|---|---|
| 7.0392 | gills (Imperial) per minute |
| 0.117320 | gills (Imperial) per second |
| 12,173.472003 | gills (U.S.) per day |
| 507.228 | gills (U.S.) per hour |
| 8.4538 | gills (U.S.) per minute |
| 0.140897 | gills (U.S.) per second |
| 14,400 | deciliters per day |
| 600 | deciliters per hour |
| 10 | deciliters per minute |
| 0.166667 | deciliters per second |
| 87,875.999971 | cubic inches per day |
| 3,661.500 | cubic inches per hour |
| 61.025 | cubic inches per minute |
| 1.0170833 | cubic inches per second |
| 144,000 | centiliters per day |
| 6,000 | centiliters per hour |
| 100 | centiliters per minute |
| 1.666667 | centiliters per second |
| 1,440,000 | milliliters per day |
| 60,000 | milliliters per hour |
| 1,000 | milliliters per minute |
| 16.666667 | milliliters per second |
| 1,440,000 | cubic centimeters per day |
| 60,000 | cubic centimeters per hour |
| 1,000 | cubic centimeters per minute |
| 16.666667 | cubic centimeters per second |
| 1,440,000,000 | cubic millimeters per day |
| 60,000,000 | cubic millimeters per hour |
| 1,000,000 | cubic millimeters per minute |
| 16,666.666667 | cubic millimeters per second |
| 48,693.167997 | ounces (U.S.) fluid per day |
| 2,028.882 | ounces (U.S.) fluid per hour |
| 33.8147 | ounces (U.S.) fluid per minute |
| 0.563578 | ounces (U.S.) fluid per second |
| 50,682.240 | ounces (Imperial—fluid) per day |
| 2,111.760 | ounces (Imperial—fluid) per hour |
| 35.196 | ounces (Imperial—fluid) per minute |
| 0.586600 | ounces (Imperial—fluid) per second |
| 389,545.776029 | drams (fluid) per day |
| 16,231.0740 | drams (fluid) per hour |
| 270.5179 | drams (fluid) per minute |
| 4.508632 | drams (fluid) per second |
| 23,372,747 | minims per day |
| 973,864.440 | minims per hour |

LITER PER MINUTE (cont'd): =

16,231.0740 . . . . . . . . . . . . . . . . . . . . . . . . .minims per minute
270.5179 . . . . . . . . . . . . . . . . . . . . . . . . . . . .minims per second

LITER PER SECOND: =

86.4 . . . . . . . . . . . . . . . . . . . . . . . . . . . . . . .kiloliters per day
3.60 . . . . . . . . . . . . . . . . . . . . . . . . . . . . . . kiloliters per hour
0.060 . . . . . . . . . . . . . . . . . . . . . . . . . . .kiloliters per minute
0.001 . . . . . . . . . . . . . . . . . . . . . . . . . . . kiloliters per second
86.4 . . . . . . . . . . . . . . . . . . . . . . . . . . cubic meters per day
3.60 . . . . . . . . . . . . . . . . . . . . . . . . . . . cubic meters per hour
0.060 . . . . . . . . . . . . . . . . . . . . . . . . cubic meters per minute
0.001 . . . . . . . . . . . . . . . . . . . . . . . . .cubic meters per second
113.0112 . . . . . . . . . . . . . . . . . . . . . . . . cubic yards per day
4.708800 . . . . . . . . . . . . . . . . . . . . . . .cubic yards per hour
0.0784800 . . . . . . . . . . . . . . . . . . . . . cubic yards per minute
0.0013080 . . . . . . . . . . . . . . . . . . . . . cubic yards per second
543.451886 . . . . . . . . . . . . . . . . . . . . . . . barrels per day
22.643829 . . . . . . . . . . . . . . . . . . . . . . . barrels per hour
0.377397 . . . . . . . . . . . . . . . . . . . . . . . . barrels per minute
0.00628995 . . . . . . . . . . . . . . . . . . . . . barrels per second
864 . . . . . . . . . . . . . . . . . . . . . . . . . . . . . .hektoliters per day
36 . . . . . . . . . . . . . . . . . . . . . . . . . . . . . . hektoliters per hour
0.60 . . . . . . . . . . . . . . . . . . . . . . . . . . .hektoliters per minute
0.01 . . . . . . . . . . . . . . . . . . . . . . . . . . .hektoliters per second
2,451.859200 . . . . . . . . . . . . . . . . . .bushels (U.S.–dry) per day
102.160800 . . . . . . . . . . . . . . . . . . .bushels (U.S.–dry) per hour
1.702680 . . . . . . . . . . . . . . . . . . . .bushels (U.S.–dry) per minute
0.028378 . . . . . . . . . . . . . . . . . . . . bushels (U.S.–dry) per second
2,375.740800 . . . . . . . . . . . . . . . . bushels (Imperial–dry) per day
98.989200 . . . . . . . . . . . . . . . . . . .bushels (Imperial–dry) per hour
1.649820 . . . . . . . . . . . . . . . . . . bushels (Imperial–dry) per minute
0.027497 . . . . . . . . . . . . . . . . . . bushels (Imperial–dry) per second
3,051.164160 . . . . . . . . . . . . . . . . . . . . . . . . cubic feet per day
127.131840 . . . . . . . . . . . . . . . . . . . . . . . cubic feet per hour
2.118864 . . . . . . . . . . . . . . . . . . . . . . . . cubic feet per minute
0.0353144 . . . . . . . . . . . . . . . . . . . . . . . cubic feet per second
8,640 . . . . . . . . . . . . . . . . . . . . . . . . . . . . dekaliters per day
360 . . . . . . . . . . . . . . . . . . . . . . . . . . . . . .dekaliters per hour
6.0 . . . . . . . . . . . . . . . . . . . . . . . . . . . . . dekaliters per minute
0.1 . . . . . . . . . . . . . . . . . . . . . . . . . . . . . dekaliters per second
9,807.436800 . . . . . . . . . . . . . . . . . . . .pecks (U.S.–dry) per day
408.643200 . . . . . . . . . . . . . . . . . . . . . pecks (U.S.–dry) per hour

LITER PER SECOND (cont'd): =

6,810720 . . . . . . . . . . . . . . . . . . . . . . . . . .pecks (U.S.—dry) per minute
0.113512 . . . . . . . . . . . . . . . . . . . . . . . . . . .pecks (U.S.—dry) per second
22,824.979200 . . . . . . . . . . . . . . . . . . . . .gallons (U.S.—liquid) per day
951.040800 . . . . . . . . . . . . . . . . . . . . . . .gallons (U.S.—liquid) per hour
15.850680 . . . . . . . . . . . . . . . . . . . . . . .gallons (U.S.—liquid) per minute
0.264178 . . . . . . . . . . . . . . . . . . . . . . . .gallons (U.S.—liquid) per second
19,614.52800 . . . . . . . . . . . . . . . . . . . . . . . . gallons (U.S.—dry) per day
817.27200 . . . . . . . . . . . . . . . . . . . . . . . . . gallons (U.S.—dry) per hour
13.62120 . . . . . . . . . . . . . . . . . . . . . . . . . gallons (U.S.—dry) per minute
0.22702 . . . . . . . . . . . . . . . . . . . . . . . . . . gallons (U.S.—dry) per second
19,006.27200 . . . . . . . . . . . . . . . . . . . . . . . . .gallons (Imperial) per day
791.92800 . . . . . . . . . . . . . . . . . . . . . . . . . gallons (Imperial) per hour
13.19880 . . . . . . . . . . . . . . . . . . . . . . . . .gallons (Imperial) per minute
0.21998 . . . . . . . . . . . . . . . . . . . . . . . . . .gallons (Imperial) per second
91,299.744 . . . . . . . . . . . . . . . . . . . . . . . . . . . quarts (liquid) per day
3,804.156 . . . . . . . . . . . . . . . . . . . . . . . . . . . quarts (liquid) per hour
63.402600 . . . . . . . . . . . . . . . . . . . . . . . . . . quarts (liquid) per minute
1.056710 . . . . . . . . . . . . . . . . . . . . . . . . . . .quarts (liquid) per second
78,460.01280 . . . . . . . . . . . . . . . . . . . . . . . . . . . . quarts (dry) per day
3,269.167200 . . . . . . . . . . . . . . . . . . . . . . . . . . . quarts (dry) per hour
54.486120 . . . . . . . . . . . . . . . . . . . . . . . . . . . . quarts (dry) per minute
0.908102 . . . . . . . . . . . . . . . . . . . . . . . . . . . . quarts (dry) per second
86,400 . . . . . . . . . . . . . . . . . . . . . . . . . . . . . . . . . . . liters per day
3,600 . . . . . . . . . . . . . . . . . . . . . . . . . . . . . . . . . . . liters per hour
60 . . . . . . . . . . . . . . . . . . . . . . . . . . . . . . . . . . . . liters per minute
1 . . . . . . . . . . . . . . . . . . . . . . . . . . . . . . . . . . . . liters per second
86,400 . . . . . . . . . . . . . . . . . . . . . . . . . . . . cubic decimeters per day
3,600 . . . . . . . . . . . . . . . . . . . . . . . . . . . . cubic decimeters per hour
60 . . . . . . . . . . . . . . . . . . . . . . . . . . . . . cubic decimeters per minute
1 . . . . . . . . . . . . . . . . . . . . . . . . . . . . . cubic decimeters per second
156,920 . . . . . . . . . . . . . . . . . . . . . . . . . . . . . pints (U.S.—dry) per day
6,538.320 . . . . . . . . . . . . . . . . . . . . . . . . . . . pints (U.S.—dry) per hour
108.9720 . . . . . . . . . . . . . . . . . . . . . . . . . . . pints (U.S.—dry) per minute
1.8162 . . . . . . . . . . . . . . . . . . . . . . . . . . . . pints (U.S.—dry) per second
182,598 . . . . . . . . . . . . . . . . . . . . . . . . . . . .pints (U.S.—liquid) per day
7,608.2400 . . . . . . . . . . . . . . . . . . . . . . . . . pints (U.S.—liquid) per hour
126.8040 . . . . . . . . . . . . . . . . . . . . . . . . .pints (U.S.—liquid) per minute
2.1134 . . . . . . . . . . . . . . . . . . . . . . . . . .pints (U.S.—liquid) per second
608,187 . . . . . . . . . . . . . . . . . . . . . . . . . . . . . gills (Imperial) per day
25,341.12 . . . . . . . . . . . . . . . . . . . . . . . . . . . .gills (Imperial) per hour
422.3520 . . . . . . . . . . . . . . . . . . . . . . . . . . gills (Imperial) per minute
7.0392 . . . . . . . . . . . . . . . . . . . . . . . . . . . gills (Imperial) per second
730,408 . . . . . . . . . . . . . . . . . . . . . . . . . . . . . . . . .gills (U.S.) per day
30,433.68 . . . . . . . . . . . . . . . . . . . . . . . . . . . . . . . gills (U.S.) per hour

**LITER PER SECOND (cont'd): =**

507.2280 . . . . . . . . . . . . . . . . . . . . . . . . . . . . . .gills (U.S.) per minute
8.4538 . . . . . . . . . . . . . . . . . . . . . . . . . . . . . .gills (U.S.) per second
864,000 . . . . . . . . . . . . . . . . . . . . . . . . . . . .deciliters per day
36,000 . . . . . . . . . . . . . . . . . . . . . . . . . . . .deciliters per hour
600 . . . . . . . . . . . . . . . . . . . . . . . . . . . .deciliters per minute
10 . . . . . . . . . . . . . . . . . . . . . . . . . . . .deciliters per second
5,272,560 . . . . . . . . . . . . . . . . . . . . . . . . .cubic inches per day
219,690 . . . . . . . . . . . . . . . . . . . . . . . . . cubic inches per hour
3,661.500 . . . . . . . . . . . . . . . . . . . . . . . .cubic inches per minute
61.025 . . . . . . . . . . . . . . . . . . . . . . . . .cubic inches per second
8,640,000 . . . . . . . . . . . . . . . . . . . . . . . . centiliters per day
360,000 . . . . . . . . . . . . . . . . . . . . . . . . centiliters per hour
6,000 . . . . . . . . . . . . . . . . . . . . . . . . . centiliters per minute
100 . . . . . . . . . . . . . . . . . . . . . . . . . centiliters per second
86,400,000 . . . . . . . . . . . . . . . . . . . . . . . milliliters per day
3,600,000 . . . . . . . . . . . . . . . . . . . . . . . .milliliters per hour
60,000 . . . . . . . . . . . . . . . . . . . . . . . . milliliters per minute
1,000 . . . . . . . . . . . . . . . . . . . . . . . . milliliters per second
86,400,000 . . . . . . . . . . . . . . . . . . . . .cubic centimeters per day
3,600,000 . . . . . . . . . . . . . . . . . . . . . cubic centimeters per hour
60,000 . . . . . . . . . . . . . . . . . . . . . .cubic centimeters per minute
1,000 . . . . . . . . . . . . . . . . . . . . . .cubic centimeters per second
86,400,000,000 . . . . . . . . . . . . . . . . . . .cubic millimeters per day
3,600,000,000 . . . . . . . . . . . . . . . . . . . cubic millimeters per hour
60,000,000 . . . . . . . . . . . . . . . . . . . . .cubic millimeters per minute
1,000,000 . . . . . . . . . . . . . . . . . . . . . cubic millimeters per second
2,921,590 . . . . . . . . . . . . . . . . . . . . .ounces (U.S.) fluid per day
121,733 . . . . . . . . . . . . . . . . . . . . . ounces (U.S.) fluid per hour
2,028.8820 . . . . . . . . . . . . . . . . . . . .ounces (U.S.) fluid per minute
33.8147 . . . . . . . . . . . . . . . . . . . . . ounces (U.S.) fluid per second
3,040,934 . . . . . . . . . . . . . . . . . . .ounces (Imperial—fluid) per day
126,706 . . . . . . . . . . . . . . . . . . . ounces (Imperial—fluid) per hour
2,111.76 . . . . . . . . . . . . . . . . . . .ounces (Imperial—fluid) per minute
35.196 . . . . . . . . . . . . . . . . . . . ounces (Imperial—fluid) per second
23,372,747 . . . . . . . . . . . . . . . . . . . . . . . drams (fluid) per day
973,864 . . . . . . . . . . . . . . . . . . . . . . . .drams (fluid) per hour
16,231.0740 . . . . . . . . . . . . . . . . . . . . . drams (fluid) per minute
270.5179 . . . . . . . . . . . . . . . . . . . . . . drams (fluid) per second
1,402,364,794 . . . . . . . . . . . . . . . . . . . . . . .minims per day
58,431,866 . . . . . . . . . . . . . . . . . . . . . . . . minims per hour
973,864 . . . . . . . . . . . . . . . . . . . . . . . . . .minims per minute
16,231.0740 . . . . . . . . . . . . . . . . . . . . . . .minims per second

METER: =

|  |  |
|---|---|
| 0.00053961 | miles (nautical) |
| 0.00062137 | miles (statute) |
| 0.001 | kilometers |
| 0.00497097 | furlongs |
| 0.01 | hektometers |
| 0.0497097 | chains |
| 0.1 | dekameters |
| 0.198839 | rods |
| 1 | meters |
| 1.093611 | yards |
| 1.811 | varas (Texas) |
| 3.280833 | feet |
| 4.374440 | spans |
| 4.970974 | links |
| 9.84250 | hands |
| 10 | decimeters |
| 100 | centimeters |
| 39.370 | inches |
| 1,000 | millimeters |
| 39,370 | mils |
| 1,000,000 | microns |
| 1,000,000,000 | milli-microns |
| 1,000,000,000 | micro-millimeters |
| 1,553,164 | wave lengths of red line of cadmium |
| 1,000,000 | Angstrom Units |
| 0.54681 | fathoms |

METER PER DAY: =

|  |  |
|---|---|
| 0.00053961 | miles (nautical) per day |
| 0.0000224837 | miles (nautical) per hour |
| 0.000000374729 | miles (nautical) per minute |
| 0.00000000624549 | miles (nautical) per second |
| 0.00062137 | miles (statute) per day |
| 0.0000258904 | miles (statute) per hour |
| 0.000000431507 | miles (statute) per minute |
| 0.00000000719178 | miles (statute) per second |
| 0.001 | kilometers per day |
| 0.0000416667 | kilometers per hour |
| 0.000000694444 | kilometers per minute |
| 0.0000000115741 | kilometers per second |
| 0.00497097 | furlongs per day |
| 0.000207124 | furlongs per hour |

METER PER DAY (cont'd): =

0.00000345206 . . . . . . . . . . . . . . . . . . . . . . . . . . furlongs per minute
0.0000000575344 . . . . . . . . . . . . . . . . . . . . . . furlongs per second
0.01 . . . . . . . . . . . . . . . . . . . . . . . . . . . . . . . . . .hektometers per day
0.000416667 . . . . . . . . . . . . . . . . . . . . . . . . . hektometers per hour
0.00000694444 . . . . . . . . . . . . . . . . . . . . . .hektometers per minute
0.000000115741 . . . . . . . . . . . . . . . . . . . . .hektometers per second
0.0497097 . . . . . . . . . . . . . . . . . . . . . . . . . . . . . . chains per day
0.00207124 . . . . . . . . . . . . . . . . . . . . . . . . . . . . .chains per hour
0.0000345206 . . . . . . . . . . . . . . . . . . . . . . . . . chains per minute
0.000000575344 . . . . . . . . . . . . . . . . . . . . . . . chains per second
0.1 . . . . . . . . . . . . . . . . . . . . . . . . . . . . . . . . . dekameters per day
0.00416667 . . . . . . . . . . . . . . . . . . . . . . . . . . .dekameters per hour
0.0000694444 . . . . . . . . . . . . . . . . . . . . . . dekameters per minute
0.00000115741 . . . . . . . . . . . . . . . . . . . . . . dekameters per second
0.198839 . . . . . . . . . . . . . . . . . . . . . . . . . . . . . . . . rods per day
0.00828496 . . . . . . . . . . . . . . . . . . . . . . . . . . . . . .rods per hour
0.000138083 . . . . . . . . . . . . . . . . . . . . . . . . . . . rods per minute
0.00000230138 . . . . . . . . . . . . . . . . . . . . . . . . . . rods per second
1 . . . . . . . . . . . . . . . . . . . . . . . . . . . . . . . . . . . . meters per day
0.0416667 . . . . . . . . . . . . . . . . . . . . . . . . . . . . . meters per hour
0.000694444 . . . . . . . . . . . . . . . . . . . . . . . . . . meters per minute
0.0000115741 . . . . . . . . . . . . . . . . . . . . . . . . . . meters per second
1.093611 . . . . . . . . . . . . . . . . . . . . . . . . . . . . . . . .yards per day
0.0455671 . . . . . . . . . . . . . . . . . . . . . . . . . . . . . . yards per hour
0.000759452 . . . . . . . . . . . . . . . . . . . . . . . . . . .yards per minute
0.0000126575 . . . . . . . . . . . . . . . . . . . . . . . . . . .yards per second
1.1811 . . . . . . . . . . . . . . . . . . . . . . . . . . . varas (Texas) per day
0.0492125 . . . . . . . . . . . . . . . . . . . . . . . . .varas (Texas) per hour
0.000820208 . . . . . . . . . . . . . . . . . . . . . varas (Texas) per minute
0.0000136701 . . . . . . . . . . . . . . . . . . . . . varas (Texas) per second
3.280833 . . . . . . . . . . . . . . . . . . . . . . . . . . . . . . . . .feet per day
0.133680 . . . . . . . . . . . . . . . . . . . . . . . . . . . . . . . . feet per hour
0.00222801 . . . . . . . . . . . . . . . . . . . . . . . . . . . . .feet per minute
0.0000371334 . . . . . . . . . . . . . . . . . . . . . . . . . . . .feet per second
4.374440 . . . . . . . . . . . . . . . . . . . . . . . . . . . . . . . .spans per day
0.182268 . . . . . . . . . . . . . . . . . . . . . . . . . . . . . . . spans per hour
0.00303781 . . . . . . . . . . . . . . . . . . . . . . . . . . . .spans per minute
0.0000506301 . . . . . . . . . . . . . . . . . . . . . . . . . . .spans per second
4.970974 . . . . . . . . . . . . . . . . . . . . . . . . . . . . . . . . links per day
0.207124 . . . . . . . . . . . . . . . . . . . . . . . . . . . . . . . links per hour
0.00345206 . . . . . . . . . . . . . . . . . . . . . . . . . . . . links per minute
0.0000575344 . . . . . . . . . . . . . . . . . . . . . . . . . . . links per second
10 . . . . . . . . . . . . . . . . . . . . . . . . . . . . . . . . . . .decimeters per day
0.416667 . . . . . . . . . . . . . . . . . . . . . . . . . . . . . decimeters per hour

METER PER DAY (cont'd): =

| | |
|---|---|
| 0.00694444 | decimeters per minute |
| 0.000115741 | decimeters per second |
| 100 | centimeters per day |
| 4.166667 | centimeters per hour |
| 0.0694444 | centimeters per minute |
| 0.00115741 | centimeters per second |
| 39.370 | inches per day |
| 1.640417 | inches per hour |
| 0.0273403 | inches per minute |
| 0.000455671 | inches per second |
| 1,000 | millimeters per day |
| 41.666667 | millimeters per hour |
| 0.694444 | millimeters per minute |
| 0.0115741 | millimeters per second |
| 39,370 | mils per day |
| 1,640.416667 | mils per hour |
| 27.340278 | mils per minute |
| 0.455671 | mils per second |
| 1,000,000 | microns per day |
| 41,666.666400 | microns per hour |
| 694.444444 | microns per minute |
| 11.574074 | microns per second |
| 1,000,000 | Angstrom Units per day |
| 41,666.666400 | Angstrom Units per hour |
| 694.444444 | Angstrom Units per minute |
| 11.574074 | Angstrom Units per second |

METER PER HOUR: =

| | |
|---|---|
| 0.0129506 | miles (nautical) per day |
| 0.00053961 | miles (nautical) per hour |
| 0.00000899350 | miles (nautical) per minute |
| 0.000000149892 | miles (nautical) per second |
| 0.0149129 | miles (statute) per day |
| 0.00062137 | miles (statute) per hour |
| 0.0000103562 | miles (statute) per minute |
| 0.000000172603 | miles (statute) per second |
| 0.0240 | kilometers per day |
| 0.001 | kilometers per hour |
| 0.0000166667 | kilometers per minute |
| 0.000000277778 | kilometers per second |
| 0.119303 | furlongs per day |
| 0.00497097 | furlongs per hour |

METER PER HOUR (cont'd): =

| | |
|---|---|
| 0.0000828495 | furlongs per minute |
| 0.00000138083 | furlongs per second |
| 0.240 | hektometers per day |
| 0.01 | hektometers per hour |
| 0.000166667 | hektometers per minute |
| 0.00000277778 | hektometers per second |
| 1.193033 | chains per day |
| 0.0497097 | chains per hour |
| 0.000828495 | chains per minute |
| 0.0000138083 | chains per second |
| 2.4 | dekameters per day |
| 0.1 | dekameters per hour |
| 0.00166667 | dekameters per minute |
| 0.0000277778 | dekameters per second |
| 4.772136 | rods per day |
| 0.198839 | rods per hour |
| 0.00331398 | rods per minute |
| 0.0000552331 | rods per second |
| 24 | meters per day |
| 1 | meter per hour |
| 0.0166667 | meters per minute |
| 0.000277778 | meters per second |
| 26.246664 | yards per day |
| 1.093611 | yards per hour |
| 0.0182268 | yards per minute |
| 0.000303781 | yards per second |
| 28.346400 | varas (Texas) per day |
| 1.1811 | varas (Texas) per hour |
| 0.0196850 | varas (Texas) per minute |
| 0.000328083 | varas (Texas) per second |
| 78.74 | feet per day |
| 3.280833 | feet per hour |
| 0.0546806 | feet per minute |
| 0.000911343 | feet per second |
| 104.986560 | spans per day |
| 4.374440 | spans per hour |
| 0.0729073 | spans per minute |
| 0.00121512 | spans per second |
| 119.303280 | links per day |
| 4.970974 | links per hour |
| 0.0828495 | links per minute |
| 0.00138083 | links per second |
| 240 | decimeters per day |
| 10 | decimeters per hour |

METER PER HOUR (cont'd): =

0.166667 . . . . . . . . . . . . . . . . . . . . . . . . . . . . .decimeters per minute
0.00277778 . . . . . . . . . . . . . . . . . . . . . . . . . .decimeters per second
2,400 . . . . . . . . . . . . . . . . . . . . . . . . . . . . . . centimeters per day
100 . . . . . . . . . . . . . . . . . . . . . . . . . . . . . . . centimeters per hour
1.666667 . . . . . . . . . . . . . . . . . . . . . . . . . . . centimeters per minute
0.0277778 . . . . . . . . . . . . . . . . . . . . . . . . . . centimeters per second
944.880 . . . . . . . . . . . . . . . . . . . . . . . . . . . . . . . . . inches per day
39.370 . . . . . . . . . . . . . . . . . . . . . . . . . . . . . . . .inches per hour
0.656167 . . . . . . . . . . . . . . . . . . . . . . . . . . . . . inches per minute
0.0109361 . . . . . . . . . . . . . . . . . . . . . . . . . . . . inches per second
24,000 . . . . . . . . . . . . . . . . . . . . . . . . . . . . . millimeters per day
1,000 . . . . . . . . . . . . . . . . . . . . . . . . . . . . . .millimeters per hour
16.666667 . . . . . . . . . . . . . . . . . . . . . . . . . . millimeters per minute
0.277778 . . . . . . . . . . . . . . . . . . . . . . . . . . . millimeters per second
944,879.999004 . . . . . . . . . . . . . . . . . . . . . . . . . mils per day
39,370 . . . . . . . . . . . . . . . . . . . . . . . . . . . . . . . . . mils per hour
656.166667 . . . . . . . . . . . . . . . . . . . . . . . . . . . mils per minute
10.936111 . . . . . . . . . . . . . . . . . . . . . . . . . . . . . .mils per second
24,000,000 . . . . . . . . . . . . . . . . . . . . . . . . . . . microns per day
1,000,000 . . . . . . . . . . . . . . . . . . . . . . . . . . . . .microns per hour
16,666.666667 . . . . . . . . . . . . . . . . . . . . . . . . microns per minute
277.777778 . . . . . . . . . . . . . . . . . . . . . . . . . . microns per second
24,000,000 . . . . . . . . . . . . . . . . . . . . . . . . . .Ansgtrom Units per day
1,000,000 . . . . . . . . . . . . . . . . . . . . . . . . . . Angstrom Units per hour
16,666.666667 . . . . . . . . . . . . . . . . . . . . . .Angstrom Units per minute
277.777778 . . . . . . . . . . . . . . . . . . . . . . . . .Angstrom Units per second

METER PER MINUTE: =

0.777038 . . . . . . . . . . . . . . . . . . . . . . . . . . . . miles (nautical) per day
0.0323766 . . . . . . . . . . . . . . . . . . . . . . . . . . miles (nautical) per hour
0.00053961 . . . . . . . . . . . . . . . . . . . . . . . .miles (nautical) per minute
0.00000899350 . . . . . . . . . . . . . . . . . . . . . miles (nautical) per second
0.894773 . . . . . . . . . . . . . . . . . . . . . . . . . . . . miles (statute) per day
0.0372822 . . . . . . . . . . . . . . . . . . . . . . . . . . . miles (statute) per hour
0.00062137 . . . . . . . . . . . . . . . . . . . . . . . . . miles (statute) per minute
0.0000103562 . . . . . . . . . . . . . . . . . . . . . . .miles (statute) per second
1.440 . . . . . . . . . . . . . . . . . . . . . . . . . . . . . . . . .kilometers per day
0.0600 . . . . . . . . . . . . . . . . . . . . . . . . . . . . . . kilometers per hour
0.001 . . . . . . . . . . . . . . . . . . . . . . . . . . . . . . .kilometers per minute
0.0000166667 . . . . . . . . . . . . . . . . . . . . . . . . kilometers per second
7.158197 . . . . . . . . . . . . . . . . . . . . . . . . . . . . . . . . furlongs per day
0.298258 . . . . . . . . . . . . . . . . . . . . . . . . . . . . . . . furlongs per hour

METER PER MINUTE (cont'd): =

| | |
|---|---|
| 0.00497097 | furlongs per minute |
| 0.0000828495 | furlongs per second |
| 14.40 | hektometers per day |
| 0.600 | hektometers per hour |
| 0.01 | hektometers per minute |
| 0.000166667 | hektometers per second |
| 71.581968 | chains per day |
| 2.982582 | chains per hour |
| 0.0497097 | chains per minute |
| 0.000828495 | chains per second |
| 144.0 | dekameters per day |
| 6.0 | dekameters per hour |
| 0.1 | dekameters per minute |
| 0.00166667 | dekameters per second |
| 286.328160 | rods per day |
| 11.930340 | rods per hour |
| 0.198839 | rods per minute |
| 0.00331398 | rods per second |
| 1,440 | meters per day |
| 60 | meters per hour |
| 1 | meters per minute |
| 0.0166667 | meters per second |
| 1,574.799840 | yards per day |
| 65.616660 | yards per hour |
| 1.093611 | yards per minute |
| 0.0182269 | yards per second |
| 1,700.784 | varas (Texas) per day |
| 70.866 | varas (Texas) per hour |
| 1.1811 | varas (Texas) per minute |
| 0.0196850 | varas (Texas) per second |
| 4,724.400 | feet per day |
| 196.85 | feet per hour |
| 3.280833 | feet per minute |
| 0.0546806 | feet per second |
| 6,299.193600 | spans per day |
| 262.466400 | spans per hour |
| 4.374440 | spans per minute |
| 0.0729073 | spans per second |
| 7,158.196800 | links per day |
| 298.258200 | links per hour |
| 4.970974 | links per minute |
| 0.0828495 | links per second |
| 14,400 | decimeters per day |
| 600 | decimeters per hour |

METER PER MINUTE (cont'd): =

| | |
|---|---|
| 10 | decimeters per minute |
| 0.166667 | decimeters per second |
| 144,000 | centimeters per day |
| 6,000 | centimeters per hour |
| 100 | centimeters per minute |
| 1.666667 | centimeters per second |
| 56,692.800 | inches per day |
| 2,362.200 | inches per hour |
| 39.370 | inches per minute |
| 0.656167 | inches per second |
| 1,440,000 | millimeters per day |
| 60,000 | millimeters per hour |
| 1,000 | millimeters per minute |
| 16.666667 | millimeters per second |
| 56,692,800 | mils per day |
| 2,362,200 | mils per hour |
| 39,370 | mils per minute |
| 656.166667 | mils per second |
| 1,440,000,000 | microns per day |
| 60,000,000 | microns per hour |
| 1,000,000 | microns per minute |
| 16,666.666667 | microns per second |
| 1,440,000,000 | Angstrom Units per day |
| 60,000,000 | Angstrom Units per hour |
| 1,000,000 | Angstrom Units per minute |
| 16,666.666667 | Angstrom Units per second |

METER PER SECOND: =

| | |
|---|---|
| 46.622304 | miles (nautical) per day |
| 1.942596 | miles (nautical) per hour |
| 0.0323766 | miles (nautical) per minute |
| 0.00053961 | miles (nautical) per second |
| 53.686368 | miles (statute) per day |
| 2.236932 | miles (statute) per hour |
| 0.0372822 | miles (statute) per minute |
| 0.00062137 | miles (statute) per second |
| 86.4 | kilometers per day |
| 3.60 | kilometers per hour |
| 0.060 | kilometers per minute |
| 0.001 | kilometers per second |
| 429.491808 | furlongs per day |
| 17.895492 | furlongs per hour |

## METER PER SECOND (cont'd): =

0.298258 . . . furlongs per minute
0.00497097 . . . furlongs per second
864 . . . hektometers per day
36 . . . hektometers per hour
0.60 . . . hektometers per minute
0.01 . . . hektometers per second
4,294.918080 . . . chains per day
178.954920 . . . chains per hour
2.982582 . . . chains per minute
0.0497097 . . . chains per second
8,640 . . . dekameters per day
360 . . . dekameters per hour
6.0 . . . dekameters per minute
0.1 . . . dekameters per second
17,179.689600 . . . rods per day
715.820400 . . . rods per hour
11.930340 . . . rods per minute
0.198839 . . . rods per second
86,400 . . . meters per day
3,600 . . . meters per hour
60 . . . meters per minute
1 . . . meters per second
94,487.990400 . . . yards per day
3,936.999600 . . . yards per hour
65.616660 . . . yards per minute
1.093611 . . . yards per second
102,047.0400 . . . varas (Texas) per day
4,251.9600 . . . varas (Texas) per hour
70.8660 . . . varas (Texas) per minute
1.1811 . . . varas (Texas) per second
283,463.971200 . . . feet per day
11,810.998800 . . . feet per hour
196.849980 . . . feet per minute
3.280833 . . . feet per second
377,951.616000 . . . spans per day
15,747.984000 . . . spans per hour
262.466400 . . . spans per minute
4.374440 . . . spans per second
429,492.153600 . . . links per day
17,895.506400 . . . links per hour
298.258440 . . . links per minute
4.970974 . . . links per second
864,000 . . . decimeters per day
36,000 . . . decimeters per hour

METER PER SECOND (cont'd): =

600 . . . . . . . . . . . . . . . . . . . . . . . . . . . . . . . .decimeters per minute
10 . . . . . . . . . . . . . . . . . . . . . . . . . . . . . . . . .decimeters per second
8,640,000 . . . . . . . . . . . . . . . . . . . . . . . . . . . centimeters per day
360,000 . . . . . . . . . . . . . . . . . . . . . . . . . . . centimeters per hour
6,000 . . . . . . . . . . . . . . . . . . . . . . . . . . . . centimeters per minute
100 . . . . . . . . . . . . . . . . . . . . . . . . . . . . . centimeters per second
3,401,568 . . . . . . . . . . . . . . . . . . . . . . . . . . . . . inches per day
141.732.0 . . . . . . . . . . . . . . . . . . . . . . . . . . . . .inches per hour
2,362.200 . . . . . . . . . . . . . . . . . . . . . . . . . . . inches per minute
39.370 . . . . . . . . . . . . . . . . . . . . . . . . . . . . inches per second
86,400,000 . . . . . . . . . . . . . . . . . . . . . . . . . millimeters per day
3,600,000 . . . . . . . . . . . . . . . . . . . . . . . . . .millimeters per hour
60,000 . . . . . . . . . . . . . . . . . . . . . . . . . . millimeters per minute
1,000 . . . . . . . . . . . . . . . . . . . . . . . . . . . millimeters per second
3,401,568,000 . . . . . . . . . . . . . . . . . . . . . . . . . . mils per day
141,732,000 . . . . . . . . . . . . . . . . . . . . . . . . . . . mils per hour
2,362,200 . . . . . . . . . . . . . . . . . . . . . . . . . . . mils per minute
39,370 . . . . . . . . . . . . . . . . . . . . . . . . . . . . .mils per second
86,400,000,000 . . . . . . . . . . . . . . . . . . . . . . microns per day
3,600,000,000 . . . . . . . . . . . . . . . . . . . . . . .microns per hour
60,000,000 . . . . . . . . . . . . . . . . . . . . . . . . microns per minute
1,000,000 . . . . . . . . . . . . . . . . . . . . . . . . . microns per second
86,400,000,000 . . . . . . . . . . . . . . . . . . . .Angstrom Units per day
3,600,000,000 . . . . . . . . . . . . . . . . . . . . Angstrom Units per hour
60,000,000 . . . . . . . . . . . . . . . . . . . . . .Angstrom Units per minute
1,000,000 . . . . . . . . . . . . . . . . . . . . . . .Angstrom Units per second

METER OF MERCURY @ 32° F.: =

0.135964 . . . . . . . . . . . . . . . . . . . . . . .hektometers of water @ 60° F.
1.359639 . . . . . . . . . . . . . . . . . . . . . . . dekameters of water @ 60° F.
13.596390 . . . . . . . . . . . . . . . . . . . . . . . . meters of water @ 60° F.
44.604005 . . . . . . . . . . . . . . . . . . . . . . . . . .feet of water @ 60° F.
535.247985 . . . . . . . . . . . . . . . . . . . . . . . . inches of water @ 60° F.
135.963901 . . . . . . . . . . . . . . . . . . . . . .decimeters of water @ 60° F.
1,359.639013 . . . . . . . . . . . . . . . . . . . . centimeters of water @ 60° F.
13,596.39013 . . . . . . . . . . . . . . . . . . . . millimeters of water @ 60° F.
0.01 . . . . . . . . . . . . . . . . . . . . . . . . hektometers of mercury @ 32° F.
0.1 . . . . . . . . . . . . . . . . . . . . . . . . . .dekameters of mercury @ 32° F.
1 . . . . . . . . . . . . . . . . . . . . . . . . . . . . . meters of mercuy @ 32° F.
3.280833 . . . . . . . . . . . . . . . . . . . . . . . . . feet of mercury @ 32° F.
39.37 . . . . . . . . . . . . . . . . . . . . . . . . . . .inches of mercury @ 32° F.
10 . . . . . . . . . . . . . . . . . . . . . . . . . . decimeters of mercury @ 32° F.

## METER OF MERCURY @ 32° F.: =

| | |
|---|---|
| 100 | centimeters of mercury @ 32° F. |
| 1,000 | millimeters of mercury @ 32° F. |
| 149,862.505 | tons per square hektometer |
| 1,498.62505 | tons per square dekameter |
| 14.986251 | tons per square meter |
| 1.392300 | tons per square foot |
| 0.00966852 | tons per square inch |
| 0.149863 | tons per square decimeter |
| 0.00149863 | tons per square centimeter |
| 0.0000149863 | tons per square millimeter |
| 135,963,901 | kilograms per square hektometer |
| 1,359,639 | kilograms per square dekameter |
| 13,596.39 | kilograms per square meter |
| 1,263.034001 | kilograms per square foot |
| 8.771085 | kilograms per square inch |
| 135.9639 | kilograms per square decimeter |
| 1.359639 | kilograms per square centimeter |
| 0.0135964 | kilograms per square millimeter |
| 299,724,991 | pounds per square hektometer |
| 2,997,249 | pounds per square dekameter |
| 29,972.49 | pounds per square meter |
| 2,784.499982 | pounds per square foot |
| 19.337009 | pounds per square inch |
| 299.7249 | pounds per square decimeter |
| 2.997249 | pounds per square centimeter |
| 0.0299725 | pounds per square millimeter |
| 13,596,390 | hektograms per square hektometer |
| 135,963.90 | hektograms per square dekameter |
| 1,359.6390 | hektograms per square meter |
| 12,630.340015 | hektograms per square foot |
| 87.711006 | hektograms per square inch |
| 13.596390 | hektograms per square decimeter |
| 0.135964 | hektograms per square centimeter |
| 0.00135964 | hektograms per square millimeter |
| 13,596,390,130 | dekagrams per square hektometer |
| 135,963,901 | dekagrams per square dekameter |
| 1,359,639 | dekagrams per square meter |
| 126,303.4 | dekagrams per square foot |
| 877.110017 | dekagrams per square inch |
| 13,596.39 | dekagrams per square decimeter |
| 135.9639 | dekagrams per square centimeter |
| 1.359639 | dekagrams per square millimeter |
| 4,795,599,897 | ounces per square hektometer |
| 47,955,998 | ounces per square dekameter |

METER OF MERCURY @ 32° F. (cont'd): =

| | |
|---|---|
| 479,560 | ounces per square meter |
| 44,552.0 | ounces per square foot |
| 309.392019 | ounces per square inch |
| 4,795.5998 | ounces per square decimeter |
| 47.955998 | ounces per square centimeter |
| 0.479560 | ounces per square millimeter |
| 135,963,901,300 | grams per square hektometer |
| 1,359,639,013 | grams per square dekameter |
| 13,596,390 | grams per square meter |
| 1,263,034 | grams per square foot |
| 8,771.111 | grams per square inch |
| 135,963.90 | grams per square decimeter |
| 1,359.6390 | grams per square centimeter |
| 13.596390 | grams per square millimeter |
| $1,359,639,013 \times 10^3$ | decigrams per square hektometer |
| 13,596,390,130 | decigrams per square dekameter |
| 135,963,900 | decigrams per square meter |
| 12,630,340 | decigrams per square foot |
| 87,711.11 | decigrams per square inch |
| 1,359,639 | decigrams per square decimeter |
| 13,596.390 | decigrams per square centimeter |
| 135.96390 | decigrams per square millimeter |
| $1,359,639,013 \times 10^4$ | centigrams per square hektometer |
| 135,963,901,300 | centigrams per square dekameter |
| 1,359,639,013 | centigrams per square meter |
| 126,303.40 | centigrams per square foot |
| 877,111.1 | centigrams per square inch |
| 13,596,390 | centigrams per square decimeter |
| 135,963.9 | centigrams per square centimeter |
| 1,359.6390 | centigrams per square millimeter |
| $1,359,639,013 \times 10^5$ | milligrams per square hektometer |
| $1,359,639,013 \times 10^3$ | milligrams per square dekameter |
| 13,596,390,130 | milligrams per square meter |
| 1,263,034 | milligrams per square foot |
| 8,771,111 | milligrams per square inch |
| 135,963,901 | milligrams per square decimeter |
| 1,359,639 | milligrams per square centimeter |
| 13,596.390 | milligrams per square millimeter |
| $133,222 \times 10^9$ | dynes per square hektometer |
| $133,222 \times 10^7$ | dynes per square dekameter |
| $133,222 \times 10^5$ | dynes per square meter |
| 1,238,591,617 | dynes per square foot |
| 8,601,331 | dynes per square inch |
| 133,222,000 | dynes per square decimeter |

## METER OF MERCURY @ 32° F. (cont'd): =

| | |
|---|---|
| 1,332,220 | dynes per square centimeter |
| 13,322.20 | dynes per square millimeter |
| 1.333300 | bars |
| 1.315801 | atmospheres |

## METER OF WATER @ 60° F.: =

| | |
|---|---|
| 0.01 | hektometers of water @ 60° F. |
| 0.1 | dekameters of water @ 60° F. |
| 1.0 | meters of water @ 60° F. |
| 3.280833 | feet of water @ 60° F. |
| 39.37 | inches of water @ 60° F. |
| 10 | decimeters of water @ 60° F. |
| 100 | centimeters of water @ 60° F. |
| 1,000 | millimeters of water @ 60° F. |
| 0.000735510 | hektometers of mercury @ 32° F. |
| 0.00735510 | dekameters of mercury @ 32° F. |
| 0.0735510 | meters of mercury @ 32° F. |
| 2.413086 | feet of mercury @ 32° F. |
| 28.957029 | inches of mercuy @ 32° F. |
| 0.735510 | decimeters of mercury @ 32° F. |
| 7.355103 | centimeters of mercury @ 32° F. |
| 73.55103 | millimeters of mercury @ 32° F. |
| 11,022.543742 | tons per square hektometer |
| 110.225437 | tons per square dekameter |
| 1.102254 | tons per square meter |
| 0.102405 | tons per square foot |
| 0.000711128 | tons per square inch |
| 0.0110225 | tons per square decimeter |
| 0.000110225 | tons per square centimeter |
| 0.00000110225 | tons per square millimeter |
| 10,000,000 | kilograms per square hektometer |
| 100,000 | kilograms per square dekameter |
| 1,000 | kilograms per square meter |
| 92.897452 | kilograms per square foot |
| 0.645121 | kilograms per square inch |
| 10 | kilograms per square decimeter |
| 0.10 | kilograms per square centimeter |
| 0.001 | kilograms per square millimeter |
| 22,045,074 | pounds per square hektometer |
| 220,450 | pounds per square dekameter |
| 2,204.5074 | pounds per square meter |
| 204.802897 | pounds per square foot |

METER OF WATER @ 60° F.: =

| | |
|---|---|
| 1.422241 | pounds per square inch |
| 22.045074 | pounds per square decimeter |
| 0.220451 | pounds per square centimeter |
| 0.00220451 | pounds per square millimeter |
| 100,000,000 | hektograms per square hektometer |
| 1,000,000 | hektograms per square dekameter |
| 10,000 | hektograms per square meter |
| 928.974717 | hektograms per square foot |
| 6.451247 | hektograms per square inch |
| 100 | hektograms per square decimeter |
| 1 | hektograms per square centimeter |
| 0.01 | hektograms per square millimeter |
| 1,000,000,000 | dekagrams per square hektometer |
| 10,000,000 | dekagrams per square dekameter |
| 100,000 | dekagrams per square meter |
| 9,289.746972 | dekagrams per square foot |
| 64.512351 | dekagrams per square inch |
| 1,000 | dekagrams per square decimeter |
| 10 | dekagrams per square centimeter |
| 0.1 | dekagrams per square millimeter |
| 352,721,381 | ounces per square hektometer |
| 3,527,213 | ounces per square dekameter |
| 35,272.13 | ounces per square meter |
| 3,276.846123 | ounces per square foot |
| 22.755860 | ounces per square inch |
| 352.7213 | ounces per square decimeter |
| 3.527213 | ounces per square centimeter |
| 0.0352721 | ounces per square millimeter |
| 10,000,000,000 | grams per square hektometer |
| 100,000,000 | grams per square dekameter |
| 1,000,000 | grams per square meter |
| 92,897.469283 | grams per square foot |
| 645.124379 | grams per square inch |
| 10,000 | grams per square decimeter |
| 100 | grams per square centimeter |
| 1 | grams per square millimeter |
| 100,000,000,000 | decigrams per square hektometer |
| 1,000,000,000 | decigrams per square dekameter |
| 10,000,000 | decigrams per square meter |
| 928,975 | decigrams per square foot |
| 6,451.243790 | decigrams per square inch |
| 100,000 | decigrams per square decimeter |
| 1,000 | decigrams per square centimeter |
| 10 | decigrams per square millimeter |

## METER OF WATER @ 60° F.: =

| | |
|---|---|
| $10 \times 10^{11}$ | centigrams per square hektometer |
| 10,000,000,000 | centigrams per square dekameter |
| 100,000,000 | centigrams per square meter |
| 9,289,747 | centigrams per square foot |
| 64,512.437904 | centigrams per square inch |
| 1,000,000 | centigrams per square decimeter |
| 10,000 | centigrams per square centimeter |
| 100 | centigrams per square millimeter |
| $10 \times 10^{12}$ | milligrams per square hektometer |
| $10 \times 10^{10}$ | milligrams per square dekameter |
| 1,000,000,000 | milligrams per square meter |
| 92,897,468 | milligrams per square foot |
| 645,124 | milligrams per square inch |
| 10,000,000 | milligrams per square decimeter |
| 100,000 | milligrams per square centimeter |
| 1,000 | milligrams per square millimeter |
| 9,798,617,182,254 | dynes per square hektometer |
| 97,986,171,823 | dynes per square dekameter |
| 979,861,718 | dynes per square meter |
| 91,099,700 | dynes per square foot |
| 632,637 | dynes per square inch |
| 9,798,617 | dynes per square decimeter |
| 97,986.171823 | dynes per square centimeter |
| 979.861718 | dynes per square millimeter |
| 0.0966778 | bars |
| 0.0967785 | atmospheres |

## MICRON: =

| | |
|---|---|
| 0.000000000621259 | miles (statute) |
| 0.000000001 | kilometers |
| 0.00000000497097 | furlongs |
| 0.00000001 | hektometers |
| 0.0000000497097 | chains |
| 0.0000001 | dekameters |
| 0.000000198839 | rods |
| 0.000001 | meters |
| 0.00000109358 | yards |
| 0.0000011811 | varas (Texas) |
| 0.0000032808 | feet |
| 0.00000437440 | spans |
| 0.00000497097 | links |
| 0.0000098425 | hands |

MICRON: =

| | |
|---|---|
| 0.00001 | decimeters |
| 0.0001 | centimeters |
| 0.00003937 | inches |
| 0.001 | millimeters |
| 0.039370 | mils |
| 1 | microns |
| 1,000 | milli-microns |
| 1,000 | micro-millimeters |
| 1.553159 | wave lengths of red line of cadmium |
| 10,000 | Angstrom Units |

MIL: =

| | |
|---|---|
| 0.00000001578 | miles (statute) |
| 0.0000000254 | kilometers |
| 0.000000126263 | furlongs |
| 0.000000254 | hektometers |
| 0.00000126263 | chains |
| 0.00000254 | dekameters |
| 0.00000505051 | rods |
| 0.0000254 | meters |
| 0.000027777 | yards |
| 0.00003 | varas (Texas) |
| 0.000083333 | feet |
| 0.000111111 | spans |
| 0.000126263 | links |
| 0.00025 | hands |
| 0.000254 | decimeters |
| 0.00254 | centimeters |
| 0.001 | inches |
| 0.0254 | millimeters |
| 1 | mils |
| 25.4 | microns |
| 25,400 | milli-microns |
| 25,400 | micro-millimeters |
| 39.450445 | wave lengths of red line of cadmium |
| 25.4 | Angstrom Units |

## MILE (STATUTE): =

| | |
|---|---|
| 0.86836 | miles (nautical) |
| 1 | miles (statute) |
| 1.60935 | kilometers |
| 8 | furlongs |
| 16.0935 | hektometers |
| 80 | chains |
| 160.935 | dekameters |
| 320 | rods |
| 1,609.35 | meters |
| 1,760 | yards |
| 1,900.8 | varas (Texas) |
| 5,280 | feet |
| 7,040 | spans |
| 8,000 | links |
| 15,840 | hands |
| 16,093.5 | decimeters |
| 160,935 | centimeters |
| 63,360 | inches |
| 1,609,350 | millimeters |
| 63,360,000 | mils |
| 1,609,344,000 | microns |
| $1{,}609{,}344 \times 10^6$ | milli-microns |
| $1{,}609{,}344 \times 10^6$ | micro-millimeter |
| 2,499,572,909 | wave lengths of red line of cadmium |
| $16{,}093{,}440 \times 10^6$ | Angstrom Units |

## MILE (STATUTE) PER DAY: =

| | |
|---|---|
| 0.86836 | miles (nautical) per day |
| 0.0361817 | miles (nautical) per hour |
| 0.000603028 | miles (nautical) per minute |
| 0.0000100505 | miles (nautical) per second |
| 1 | miles (statute) per day |
| 0.0416667 | miles (statute) per hour |
| 0.000694444 | miles (statute) per minute |
| 0.0000115741 | miles (statute) per second |
| 1.60935 | kilometers per day |
| 0.0670562 | kilometers per hour |
| 0.00111760 | kilometers per minute |
| 0.0000186267 | kilometers per second |
| 8 | furlongs per day |
| 0.333333 | furlongs per hour |
| 0.00555556 | furlongs per minute |

MILE (STATUTE) PER DAY (cont'd): =

| | |
|---|---|
| 0.0000925926 | furlongs per second |
| 16.0935 | hektometers per day |
| 0.670562 | hektometers per hour |
| 0.0111760 | hektometers per minute |
| 0.000186267 | hektometers per second |
| 80 | chains per day |
| 3.333333 | chains per hour |
| 0.0555556 | chains per minute |
| 0.000925926 | chains per second |
| 160.935 | dekameters per day |
| 6.705625 | dekameters per hour |
| 0.111760 | dekameters per minute |
| 0.00186267 | dekameters per second |
| 320 | rods per day |
| 13.333333 | rods per hour |
| 0.222222 | rods per minute |
| 0.00370370 | rods per second |
| 1,609.35 | meters per day |
| 67.056250 | meters per hour |
| 1.117604 | meters per minute |
| 0.0186267 | meters per second |
| 1,760 | yards per day |
| 73.333333 | yards per hour |
| 1.222222 | yards per minute |
| 0.0203703 | yards per second |
| 1,900.8 | varas (Texas) per day |
| 79.2 | varas (Texas) per hour |
| 1.320 | varas (Texas) per minute |
| 0.0220 | varas (Texas) per second |
| 5,280 | feet per day |
| 220 | feet per hour |
| 3.666667 | feet per minute |
| 0.06111111 | feet per second |
| 7,040 | spans per day |
| 293.333333 | spans per hour |
| 4.888889 | spans per minute |
| 0.0814814 | spans per second |
| 8,000 | links per day |
| 333.333333 | links per hour |
| 5.555556 | links per minute |
| 0.0925925 | links per second |
| 15,840 | hands per day |
| 660 | hands per hour |
| 11 | hands per minute |

## MILE (STATUTE) PER DAY (cont'd): =

| | |
|---|---|
| 0.183333 | hands per second |
| 16,093.5 | decimeters per day |
| 670.56250 | decimeters per hour |
| 11.176042 | decimeters per minute |
| 0.186267 | decimeters per second |
| 160,935 | centimeters per day |
| 6,705.624996 | centimeters per hour |
| 111.760417 | centimeters per minute |
| 1.862674 | centimeters per second |
| 63,360 | inches per day |
| 2,640 | inches per hour |
| 44 | inches per minute |
| 0.733333 | inches per second |
| 1,609,350 | millimeters per day |
| 67,056.249960 | millimeters per hour |
| 1,117.604166 | millimeters per minute |
| 18.626736 | millimeters per second |

## MILE (STATUTE) PER HOUR: =

| | |
|---|---|
| 20.84064 | miles (nautical) per day |
| 0.86836 | miles (nautical) per hour |
| 0.0144727 | miles (nautical) per minute |
| 0.000241211 | miles (nautical) per second |
| 24 | miles (statute) per day |
| 1 | miles (statute) per hour |
| 0.0166667 | miles (statute) per minute |
| 0.000277778 | miles (statute) per second |
| 38.6244 | kilometers per day |
| 1.60935 | kilometers per hour |
| 0.0268225 | kilometers per minute |
| 0.000447042 | kilometers per second |
| 192 | furlongs per day |
| 8 | furlongs per hour |
| 0.133333 | furlongs per minute |
| 0.00222222 | furlongs per second |
| 386.244 | hektometers per day |
| 16.0935 | hektometers per hour |
| 0.268225 | hektometers per minute |
| 0.00447042 | hektometers per second |
| 1,920 | chains per day |
| 80 | chains per hour |
| 1.333333 | chains per minute |

MILE (STATUTE) PER HOUR (cont'd): =

| | |
|---|---|
| 0.0222222 | chains per second |
| 3,862.44 | dekameters per day |
| 160.935 | dekameters per hour |
| 2.682250 | dekameters per minute |
| 0.0447042 | dekameters per second |
| 7,680 | rods per day |
| 320 | rods per hour |
| 5.333333 | rods per minute |
| 0.0888889 | rods per second |
| 38,624.4 | meters per day |
| 1,609.35 | meters per hour |
| 26.8225 | meters per minute |
| 0.447042 | meters per second |
| 42,240 | yards per day |
| 1,760 | yards per hour |
| 29.333333 | yards per minute |
| 0.488889 | yards per second |
| 45,619.2 | varas (Texas) per day |
| 1,900.8 | varas (Texas) per hour |
| 31.68 | varas (Texas) per minute |
| 0.528 | varas (Texas) per second |
| 126,720 | feet per day |
| 5,280 | feet per hour |
| 88 | feet per minute |
| 1.466667 | feet per second |
| 168,960 | spans per day |
| 7,040 | spans per hour |
| 117.333333 | spans per minute |
| 1.955556 | spans per second |
| 192,000 | links per day |
| 8,000 | links per hour |
| 133.333333 | links per minute |
| 2.222222 | links per second |
| 380,160 | hands per day |
| 15,840 | hands per hour |
| 264 | hands per minute |
| 4.40 | hands per second |
| 386,244 | decimeters per day |
| 16,093.5 | decimeters per hour |
| 268.225 | decimeters per minute |
| 4.470417 | decimeters per second |
| 3,862,440 | centimeters per day |
| 160,935 | centimeters per hour |
| 2,682.25 | centimeters per minute |

## MILE (STATUTE) PER HOUR (cont'd): =

44.704167 . . . . . . . . . . centimeters per second
1,520,640 . . . . . . . . . . inches per day
63,360 . . . . . . . . . . inches per hour
1,056 . . . . . . . . . . inches per minute
17.6 . . . . . . . . . . inches per second
38,624,400 . . . . . . . . . . millimeters per day
1,609,350 . . . . . . . . . . millimeters per hour
26,822.5 . . . . . . . . . . millimeters per minute
447.0416667 . . . . . . . . . . millimeters per second

## MILE (STATUTE) PER MINUTE: =

1,250.438400 . . . . . . . . . . miles (nautical) per day
52.101600 . . . . . . . . . . miles (nautical) per hour
0.86836 . . . . . . . . . . miles (nautical) per minute
0.0144727 . . . . . . . . . . miles (nautical) per second
1,440 . . . . . . . . . . miles (statute) per day
60 . . . . . . . . . . miles (statute) per hour
1 . . . . . . . . . . miles (statute) per minute
0.01666667 . . . . . . . . . . miles (statute) per second
2,317.464 . . . . . . . . . . kilometers per day
96.561 . . . . . . . . . . kilometers per hour
1.60935 . . . . . . . . . . kilometers per minute
0.0268225 . . . . . . . . . . kilometers per second
11,520 . . . . . . . . . . furlongs per day
480 . . . . . . . . . . furlongs per hour
8 . . . . . . . . . . furlongs per minute
0.133333 . . . . . . . . . . furlongs per second
23,174.64 . . . . . . . . . . hektometers per day
965.61 . . . . . . . . . . hektometers per hour
16.0935 . . . . . . . . . . hektometers per minute
0.268225 . . . . . . . . . . hektometers per second
115,200 . . . . . . . . . . chains per day
4,800 . . . . . . . . . . chains per hour
80 . . . . . . . . . . chains per minute
1.333333 . . . . . . . . . . chains per second
231,746 . . . . . . . . . . dekameters per day
9,656.1 . . . . . . . . . . dekameters per hour
160.935 . . . . . . . . . . dekameters per minute
2.682250 . . . . . . . . . . dekameters per second
460,800 . . . . . . . . . . rods per day
19,200 . . . . . . . . . . rods per hour
320 . . . . . . . . . . rods per minute

## MILE (STATUTE) PER MINUTE (cont'd)' =

| | |
|---|---|
| 5.333333 | rods per second |
| 2,317,464 | meters per day |
| 96,561 | meters per hour |
| 1,609.35 | meters per minute |
| 26.82250 | meters per second |
| 2,534,400 | yards per second |
| 105,600 | yards per hour |
| 1,760 | yards per minute |
| 29.333333 | yards per second |
| 2,737,152 | varas (Texas) per day |
| 114,048 | varas (Texas) per hour |
| 1,900.8 | varas (Texas) per minute |
| 31.680 | varas (Texas) per second |
| 7,603,200 | feet per day |
| 316,800 | feet per hour |
| 5,280 | feet per minute |
| 88 | feet per second |
| 10,137,600 | spans per day |
| 422,400 | spans per hour |
| 7,040 | spans per minute |
| 117.333333 | spans per second |
| 11,520,000 | links per day |
| 480,000 | links per hour |
| 8,000 | links per minute |
| 133.333333 | links per second |
| 22,809,600 | hands per day |
| 950,400 | hands per hour |
| 15,840 | hands per minute |
| 264.0 | hands per second |
| 23,174,640 | decimeters per day |
| 965,610 | decimeters per hour |
| 16,093.5 | decimeters per minute |
| 268.225 | decimeters per second |
| 231,746,400 | centimeters per day |
| 9,656,100 | centimeters per hour |
| 160,935 | centimeters per minute |
| 2,682.25 | centimeters per second |
| 91,238,400 | inches per day |
| 3,801,600 | inches per hour |
| 63,360 | inches per minute |
| 1,056.0 | inches per second |
| 2,317,464,000 | millimeters per day |
| 96,561,000 | millimeters per hour |
| 1,609,350 | millimeters per minute |
| 26,822.5 | millimeters per second |

MILE (STATUTE) PER SECOND: =

| | |
|---|---|
| 75,026.304 | miles (nautical) per day |
| 3,126.096 | miles (nautical) per hour |
| 52.10260 | miles (nautical) per minute |
| 0.86836 | miles (nautical) per second |
| 86,400 | miles (statute) per day |
| 3,600 | miles (statute) per hour |
| 60 | miles (statute) per minute |
| 1 | miles (statute) per second |
| 139,048 | kilometers per day |
| 5,793.660 | kilometers per hour |
| 96.561 | kilometers per minute |
| 1.60935 | kilometers per second |
| 691,200 | furlongs per day |
| 28,800 | furlongs per hour |
| 480 | furlongs per minute |
| 8 | furlongs per second |
| 1,390,480 | hektometers per day |
| 57,936.6 | hektometers per hour |
| 965.61 | hektometers per minute |
| 16.0935 | hektometers per second |
| 6,912,000 | chains per day |
| 288,000 | chains per hour |
| 4,800 | chains per minute |
| 80 | chains per second |
| 13,904,800 | dekameters per day |
| 579,366 | dekameters per hour |
| 9,656.1 | dekameters per minute |
| 160.935 | dekameters per second |
| 27,648,000 | rods per day |
| 1,152,000 | rods per hour |
| 19,200 | rods per minute |
| 320 | rods per second |
| 139,048,000 | meters per day |
| 5,793,660 | meters per hour |
| 96,561.0 | meters per minute |
| 1,609.35 | meters per second |
| 152,064,000 | yards per day |
| 6,336,000 | yards per hour |
| 105,600 | yards per minute |
| 1,760 | yards per second |
| 164,229,120 | varas (Texas) per day |
| 6,842,880 | varas (Texas) per hour |
| 114,048 | varas (Texas) per minute |
| 1,900.8 | varas (Texas) per second |

MILE (STATUTE) PER SECOND (cont'd): =

| | |
|---|---|
| 456,192,000 | feet per day |
| 19,008,000 | feet per hour |
| 316,800 | feet per minute |
| 5,280 | feet per second |
| 608,256,000 | spans per day |
| 25,344,000 | spans per hour |
| 422,400 | spans per minute |
| 7,040 | spans per second |
| 691,200,000 | links per day |
| 28,800,000 | links per hour |
| 480,000 | links per minute |
| 8,000 | links per second |
| 1,390,478,000 | decimeters per day |
| 57,936,600 | decimeters per hour |
| 965,610 | decimeters per minute |
| 16,093.5 | decimeters per second |
| 13,904,780,000 | centimeters per day |
| 579,366,000 | centimeters per hour |
| 9,656,100 | centimeters per minute |
| 160,935 | centimeters per second |
| 5,474,304,000 | inches per day |
| 228,096,000 | inches per hour |
| 3,801,600 | inches per minute |
| 63,360 | inches per second |
| 139,047,800,000 | millimeters per day |
| 5,793,660,000 | millimeters per hour |
| 96,561,000 | millimeters per minute |
| 1,609,350 | millimeters per second |

MIL: =

| | |
|---|---|
| 0.0000000137061 | miles (nautical) |
| 0.00000001578 | miles (statute) |
| 0.0000000254 | kilometers |
| 0.000000126263 | furlongs |
| 0.000000254 | hektometers |
| 0.00000126263 | chains |
| 0.00000254 | dekameters |
| 0.00000505052 | rods |
| 0.0000254 | meters |
| 0.0000277778 | yards |
| 0.00003 | varas (Texas) |
| 0.0000833333 | feet |

MIL (cont'd): =

| | |
|---|---|
| 0.000111111 | spans |
| 0.000126263 | links |
| 0.000250 | hands |
| 0.000254 | decimeters |
| 0.00254 | centimeters |
| 0.001 | inches |
| 0.0254 | millimeters |
| 1 | mils |
| 25.4 | microns |
| 25,400 | milli-microns |
| 25,400 | micro-millimeters |
| 39.450445 | wave lengths of red line of cadmium |
| 25.4 | Angstrom Units |

MILLIMETER: =

| | |
|---|---|
| 0.00000053961 | miles (nautical) |
| 0.00000062137 | miles (statute) |
| 0.000001 | kilometers |
| 0.00000497097 | furlongs |
| 0.00001 | hektometers |
| 0.0000497097 | chains |
| 0.0001 | dekameters |
| 0.000198839 | rods |
| 0.001 | meters |
| 0.00109361 | yards |
| 0.0011811 | varas (Texas) |
| 0.00328083 | feet |
| 0.00437444 | spans |
| 0.00497097 | links |
| 0.00984250 | hands |
| 0.01 | decimeters |
| 0.1 | centimeters |
| 0.039370 | inches |
| 1 | millimeters |
| 39.370 | mils |
| 1,000 | microns |
| 1,000,000 | milli-microns |
| 1,000,000 | micro-millimeters |
| 1,553.164 | wave lengths of red line of cadmium |
| 1,000 | Angstrom Units |

MINUTE (ANGLE): =

0.00018519 . . . . . . . . . . . . . . . . . . . . . . . . . . . . . . quadrants
0.000290888 . . . . . . . . . . . . . . . . . . . . . . . . . . . . . radians
0.0166667 . . . . . . . . . . . . . . . . . . . . . . . . . . . . . . degrees
60 . . . . . . . . . . . . . . . . . . . . . . . . . . . . . . . . . . seconds
0.0000462963 . . . . . . . . . . . . . . . . . . circumferences or revolutions

MILLIGRAM: =

0.00000000098426 . . . . . . . . . . . . . . . . . . . . . . tons (long)
0.000000001 . . . . . . . . . . . . . . . . . . . . . . . . . tons (metric)
0.00000000110231 . . . . . . . . . . . . . . . . . . . . . . tons (short)
0.000001 . . . . . . . . . . . . . . . . . . . . . . . . . . . kilograms
0.00000267923 . . . . . . . . . . . . . . . . . . . . . . pounds (Troy)
0.00000220462 . . . . . . . . . . . . . . . . . . . . . pounds (Avoir.)
0.00001 . . . . . . . . . . . . . . . . . . . . . . . . . . . hektograms
0.0001 . . . . . . . . . . . . . . . . . . . . . . . . . . . . dekagrams
0.0000321507 . . . . . . . . . . . . . . . . . . . . . . ounces (Troy)
0.0000352739 . . . . . . . . . . . . . . . . . . . . . ounces (Avoir.)
0.001 . . . . . . . . . . . . . . . . . . . . . . . . . . . . . . grams
0.01 . . . . . . . . . . . . . . . . . . . . . . . . . . . . . decigrams
0.1 . . . . . . . . . . . . . . . . . . . . . . . . . . . . . centigrams
1 . . . . . . . . . . . . . . . . . . . . . . . . . . . . . . . milligrams
0.0154324 . . . . . . . . . . . . . . . . . . . . . . . . . . . . . grains
0.000257206 . . . . . . . . . . . . . . . . . . . . . . . . drams (Troy)
0.000564383 . . . . . . . . . . . . . . . . . . . . . . . drams (Avoir.)
0.000643015 . . . . . . . . . . . . . . . . . . . . . . . pennyweights
0.000771618 . . . . . . . . . . . . . . . . . . . . . . . . . . scruples
0.005 . . . . . . . . . . . . . . . . . . . . . . . . . . . carats (metric)

MILLIGRAM PER LITER: =

1 . . . . . . . . . . . . . . . . . . . . . . . . . . . . . . parts per million

MILLILITER: =

0.000001 . . . . . . . . . . . . . . . . . . . . . . . . . . . . . kiloliters
0.000001 . . . . . . . . . . . . . . . . . . . . . . . . . . . cubic meters
0.0000013080 . . . . . . . . . . . . . . . . . . . . . . . . . . cubic yards
0.00000628995 . . . . . . . . . . . . . . . . . . . . . . . . . . . . barrels
0.00001 . . . . . . . . . . . . . . . . . . . . . . . . . . . . . hektoliters
0.000028378 . . . . . . . . . . . . . . . . . . . . . . . bushels (U.S.—dry)

MILLILITER (cont'd): =

0.000027497 . . . . . . . . . . . . . . . . . . . . . . . . . . . bushels (Imperial—dry)
0.000353144 . . . . . . . . . . . . . . . . . . . . . . . . . . . . . . . . cubic feet
0.0001 . . . . . . . . . . . . . . . . . . . . . . . . . . . . . . . . dekaliters
0.000113512 . . . . . . . . . . . . . . . . . . . . . . . . . . pecks (U.S.—dry)
0.000264178 . . . . . . . . . . . . . . . . . . . . . . . . gallons (U.S.—liquid)
0.00022702 . . . . . . . . . . . . . . . . . . . . . . . . . .gallons (U.S.—dry)
0.00021998 . . . . . . . . . . . . . . . . . . . . . . . . . . gallons (Imperial)
0.00105671 . . . . . . . . . . . . . . . . . . . . . . . . . . . . quarts (liquid)
0.000908102 . . . . . . . . . . . . . . . . . . . . . . . . . . . . . .quarts (dry)
0.001 . . . . . . . . . . . . . . . . . . . . . . . . . . . . . . . . . . . .liters
0.001 . . . . . . . . . . . . . . . . . . . . . . . . . . . . . . cubic decimeters
0.0018162 . . . . . . . . . . . . . . . . . . . . . . . . . . . . . pints (U.S.—dry)
0.0021134 . . . . . . . . . . . . . . . . . . . . . . . . . . pints (U.S.—liquid)
0.0073092 . . . . . . . . . . . . . . . . . . . . . . . . . . . . . gills (Imperial)
0.0084538 . . . . . . . . . . . . . . . . . . . . . . . . . . . . . . . . gills (U.S.)
0.01 . . . . . . . . . . . . . . . . . . . . . . . . . . . . . . . . . . . deciliters
0.0610234 . . . . . . . . . . . . . . . . . . . . . . . . . . . . . . . cubic inches
0.1 . . . . . . . . . . . . . . . . . . . . . . . . . . . . . . . . . . . centiliters
1 . . . . . . . . . . . . . . . . . . . . . . . . . . . . . . . . . . . . . milliliters
1 . . . . . . . . . . . . . . . . . . . . . . . . . . . . . . . cubic centimeters
1,000 . . . . . . . . . . . . . . . . . . . . . . . . . . . . . .cubic millimeters
0.0338147 . . . . . . . . . . . . . . . . . . . . . . . . . . . ounces (U.S.—fluid)
0.035196 . . . . . . . . . . . . . . . . . . . . . . . . . .ounces (Imperial—fluid)
0.270518 . . . . . . . . . . . . . . . . . . . . . . . . . . . . . . . drams (fluid)
16.231074 . . . . . . . . . . . . . . . . . . . . . . . . . . . . . . . . . . minims
0.00220462 . . . . . . . . . . . . . . . . . . . pounds of water @ maximum density

OHM (ABSOLUTE): =

0.00000000000111263 . . . . . . . . . . . . . elestrostatic cgs unit or statohm
0.000001 . . . . . . . . . . . . . . . . . . . . . . . . . . . . megohm (absolute)
0.99948 . . . . . . . . . . . . . . . . . . . . . . . . . . . . . International ohm
1,000,000 . . . . . . . . . . . . . . . . . . . . . . . . . . . .microhms (absolute)
1,000,000,000 . . . . . . . . . . . . . . . . . . .elestromagnetic cgs units or abohms

ohm per kilometer = 0.3048 ohms per 1,000 feet
ohm per 1,000 feet = 3.280833 ohms per kilometer
ohm per 1,000 yards = 1.0936 ohms per kilometer

OUNCE (AVOIRDUPOIS): =

0.0000279018 . . . . . . . . . . . . . . . . . . . . . . . . . . . . . . tons (long)
0.0000283495 . . . . . . . . . . . . . . . . . . . . . . . . . . . . . tons (metric)

**OUNCE (AVOIRDUPOIS) (cont'd): =**

| | |
|---|---|
| 0.00003125 | tons (net) |
| 0.0282495 | kilograms |
| 0.0759549 | pounds (Troy) |
| 0.0625 | pounds (Avoir.) |
| 0.283495 | hektograms |
| 2.834953 | dekagrams |
| 0.9114583 | ounces (Troy) |
| 1 | ounces (Avoir.) |
| 28.349527 | grams |
| 283.49527 | decigrams |
| 2,834.9527 | centigrams |
| 28,349.527 | milligrams |
| 437.5 | grains |
| 7.29166 | drams (Troy) |
| 16 | drams (Avoir.) |
| 18.22917 | pennyweights |
| 21.875 | scruples |
| 141.75 | carats (metric) |

**OUNCE (FLUID): =**

| | |
|---|---|
| 0.0000295729 | kiloliters |
| 0.0000295729 | cubic meters |
| 0.0000386814 | cubic yards |
| 0.000186012 | barrels |
| 0.000295729 | hektoliters |
| 0.000839221 | bushels (U.S.—dry) |
| 0.000813167 | bushels (Imperial—dry) |
| 0.00104435 | cubic feet |
| 0.00295729 | dekaliters |
| 0.00335688 | pecks (U.S.—dry) |
| 0.00781252 | gallons (U.S.—liquid) |
| 0.00671365 | gallons (U.S.—dry) |
| 0.00650545 | gallons (Imperial) |
| 0.03125 | quarts (liquid) |
| 0.0268552 | quarts (dry) |
| 0.0295729 | liters |
| 0.0295729 | cubic decimeters |
| 0.0537104 | pints (U.S.—dry) |
| 0.0625 | pints (U.S.—liquid) |
| 0.208170 | gills (Imperial) |
| 0.25 | gills (U.S.) |
| 0.295729 | deciliters |

OUNCE (FLUID) (cont'd): =

| | |
|---|---|
| 1.80469 | cubic inches |
| 2.957294 | centiliters |
| 29.572937 | milliliters |
| 29.572937 | cubic centimeters |
| 29,572.9372 | cubic millimeters |
| 1 | ounces (U.S.—fluid) |
| 1.0408491 | ounces (Imperial—fluid) |
| 8 | drams (fluid) |
| 480 | minims |
| 0.0651972 | pounds of water — maximum density |

OUNCE (TROY): =

| | |
|---|---|
| 0.0000306122 | tons (long) |
| 0.0000311034 | tons (metric) |
| 0.000034285 | tons (net) |
| 0.0311035 | kilograms |
| 0.0833333 | pounds (Troy) |
| 0.0685714 | pounds (Avoir.) |
| 0.311035 | hektograms |
| 3.110348 | dekagrams |
| 1 | ounces (Troy) |
| 1.09714 | ounces (Avoir.) |
| 31.103481 | grams |
| 311.03481 | decigrams |
| 3,110.3481 | centigrams |
| 31,103.481 | milligrams |
| 480 | grains |
| 8 | drams (Troy) |
| 17.55428 | drams (Avoir.) |
| 20 | pennyweights |
| 24 | scruples |
| 155.52 | carats (metric) |

OUNCE (WEIGHT) PER SQUARE INCH: =

| | |
|---|---|
| 0.000439419 | hektometers of water @ 60° F. |
| 0.00439419 | dekameters of water @ 60° F. |
| 0.0439419 | meters of water @ 60° F. |
| 0.144174 | feet of water @ 60° F. |
| 1.730092 | inches of water @ 60° F. |

## OUNCE (WEIGHT) PER SQUARE INCH (cont'd): =

| | |
|---|---|
| 0.439419 | decimeters of water @ 60° F. |
| 4.394188 | centimeters of water @ 60° F. |
| 43.941875 | millimeters of water @ 60° F. |
| 0.0000323219 | hektometers of mercury @ 32° F. |
| 0.000323219 | dekameters of mercury @ 32° F. |
| 0.00323219 | meters of mercuy @ 32° F. |
| 0.0106042 | feet of mercury @ 32° F. |
| 0.127250 | inches of mercury @ 32° F. |
| 0.0323219 | decimeters of mercury @ 32° F. |
| 0.323219 | centimeters of mercury @ 32° F. |
| 3.232188 | millimeters of mercury @ 32° F. |
| 484.379356 | tons per square hektometer |
| 4.843794 | tons per square dekameter |
| 0.0484379 | tons per square meter |
| 0.00450014 | tons per square foot |
| 0.0000312500 | tons per square inch |
| 0.000484379 | tons per square decimeter |
| 0.00000484379 | tons per square centimeter |
| 0.0000000484379 | tons per square millimeter |
| 439,419 | kilograms per square hektometer |
| 4,394.1875 | kilograms per square dekameter |
| 43.941875 | kilograms per square meter |
| 4.0823252 | kilograms per square foot |
| 0.0283494 | kilograms per square inch |
| 0.439419 | kilograms per square decimeter |
| 0.00439419 | kilograms per square centimeter |
| 0.0000439419 | kilograms per square millimeter |
| 968,758 | pounds per square hektometer |
| 9,687.58 | pounds per square dekameter |
| 96.8758 | pounds per square meter |
| 90 | pounds per square foot |
| 0.0625 | pounds per square inch |
| 0.968758 | pounds per square decimeter |
| 0.00968758 | pounds per square centimeter |
| 0.0000968758 | pounds per square millimeter |
| 4,394,190 | hektograms per square hektometer |
| 43,941.875 | hektograms per square dekameter |
| 439.41875 | hektograms per square meter |
| 40.823252 | hektograms per square foot |
| 0.283494 | hektograms per square inch |
| 4.394188 | hektograms per square decimeter |
| 0.043919 | hektograms per square centimeter |
| 0.000439419 | hektograms per square millimeter |
| 43,941,900 | dekagrams per square hektometer |

OUNCE (WEIGHT) PER SQUARE INCH (cont'd): =

| | |
|---|---|
| 439,419 | dekagrams per square dekameter |
| 4,394.1875 | dekagrams per square meter |
| 408.232519 | dekagrams per square foot |
| 2.834944 | dekagrams per square inch |
| 43.941875 | dekagrams per square decimeter |
| 0.439419 | dekagrams per square centimeter |
| 0.00439419 | dekagrams per square millimeter |
| 15,500,139 | ounces per square hektometer |
| 155,001 | ounces per square dekameter |
| 1,550.0139 | ounces per square meter |
| 144 | ounces per square foot |
| 1 | ounces per square inch |
| 15.500139 | ounces per square decimeter |
| 0.155001 | ounces per square centimeter |
| 0.00155001 | ounces per square millimeter |
| 439.419,000 | grams per square hektometer |
| 4,394,190 | grams per square dekameter |
| 43,941.875 | grams per square meter |
| 4,082.325187 | grams per square foot |
| 28.349438 | grams per square inch |
| 439.4187 | grams per square decimeter |
| 4.394187 | grams per square centimeter |
| 0.0439419 | grams per square millimeter |
| 4,394,190,000 | decigrams per square hektometer |
| 43,941,900 | decigrams per square dekameter |
| 439,419 | decigrams per square meter |
| 40,823.25187 | decigrams per square foot |
| 283.494375 | decigrams per square inch |
| 4,394.1875 | decigrams per square decimeter |
| 43.94187 | decigrams per square centimeter |
| 0.439419 | decigrams per square millimeter |
| 43,941,900,000 | centigrams per square hektometer |
| 439,419,000 | centigrams per square dekameter |
| 4,394,190 | centigrams per square meter |
| 408,233 | centigrams per square foot |
| 2,834.943750 | centigrams per square inch |
| 43,941.875 | centigrams per square decimeter |
| 439.41875 | cengigrams per square centimeter |
| 4.394188 | centigrams per square millimeter |
| 439,419,000,000 | milligrams per square hektometer |
| 4,394,190,000 | milligrams per square dekameter |
| 43,941,900 | milligrams per square meter |
| 4,082,325 | milligrams per square foot |
| 28,349.43750 | milligrams per square inch |

## OUNCE (WEIGHT) PER SQUARE INCH (cont'd): =

439,419 . . . . . . . . . . . . . . . . . . . . . . . milligrams per square decimeter
4,394.1875 . . . . . . . . . . . . . . . . . . . . .milligrams per square centimeter
43.941875 . . . . . . . . . . . . . . . . . . . . . . milligrams per square millimeter
430,920,000,000 . . . . . . . . . . . . . . . . . .dynes per square hektometer
4,309,200,000 . . . . . . . . . . . . . . . . . . . . dynes per square dekameter
43,092,000 . . . . . . . . . . . . . . . . . . . . . . . . dynes per square meter
4,003,324 . . . . . . . . . . . . . . . . . . . . . . . . . dynes per square foot
28,050.875 . . . . . . . . . . . . . . . . . . . . . . . . dynes per square inch
430,920 . . . . . . . . . . . . . . . . . . . . . . . . .dynes per square decimeter
4,309.2 . . . . . . . . . . . . . . . . . . . . . . . . dynes per square centimeter
43.092 . . . . . . . . . . . . . . . . . . . . . . . . . dynes per square millimeter
0.00430919 . . . . . . . . . . . . . . . . . . . . . . . . . . . . . . . . . . . bars
0.00425288 . . . . . . . . . . . . . . . . . . . . . . . . . . . . . . atmospheres

## PARTS PER MILLION: =

0.00000110231 . . . . . . . . . . . . . . . . . . . . . . . tons (net) per cubic meter
0.001 . . . . . . . . . . . . . . . . . . . . . . . . . . . . . kilograms per cubic meter
0.00220462 . . . . . . . . . . . . . . . . . . . . . .pounds (Avoir.) per cubic meter
0.0352739 . . . . . . . . . . . . . . . . . . . . . . . ounces (Avoir.) per cubic meter
1.0 . . . . . . . . . . . . . . . . . . . . . . . . . . . . . . .grams per cubic meter
15.4324 . . . . . . . . . . . . . . . . . . . . . . . . . . . .grains per cubic meter
0.000000175250 . . . . . . . . . . . . . . . . . . . . . . . . .tons (net) per barrel
0.000158984 . . . . . . . . . . . . . . . . . . . . . . . . . . . kilograms per barrel
0.000350499 . . . . . . . . . . . . . . . . . . . . . . . . pounds (Avoir.) per barrel
0.00560799 . . . . . . . . . . . . . . . . . . . . . . . . .ounces (Avoir.) per barrel
0.158984 . . . . . . . . . . . . . . . . . . . . . . . . . . . . . . . . grams per barrel
2.453505 . . . . . . . . . . . . . . . . . . . . . . . . . . . . . . . . grains per barrel
0.0000000312133 . . . . . . . . . . . . . . . . . . . . . . . tons (net) per cubic foot
0.0000283162 . . . . . . . . . . . . . . . . . . . . . . . . . kilograms per cubic foot
0.0000624264 . . . . . . . . . . . . . . . . . . . . . . pounds (Avoir.) per cubi foot
0.000998823 . . . . . . . . . . . . . . . . . . . . . . . ounces (Avoir.) per cubic foot
0.0283162 . . . . . . . . . . . . . . . . . . . . . . . . . . . . .grams per cubic foot
0.436987 . . . . . . . . . . . . . . . . . . . . . . . . . . . . . .grains per cubic foot
0.00000000417262 . . . . . . . . . . . . . . . . . . tons (net) per gallon (U.S.—liquid)
0.00000378524 . . . . . . . . . . . . . . . . . . . . .kilograms per gallon (U.S.—liquid)
0.00000834522 . . . . . . . . . . . . . . . . . pounds (Avoir.) per gallon (U.S.—liquid)
0.000133524 . . . . . . . . . . . . . . . . . . . ounces (Avoir.) per gallon (U.S.—liquid)
0.00378534 . . . . . . . . . . . . . . . . . . . . . . . . grams per gallon (U.S.—liquid)
0.0584168 . . . . . . . . . . . . . . . . . . . . . . . . grains per gallons (U.S.—liquid)
0.00000000501107 . . . . . . . . . . . . . . . . . tons (net) per gallon (Imperial—liquid)
0.00000454585 . . . . . . . . . . . . . . . . . . kilograms per gallon (Imperial—liquid)
0.0000100221 . . . . . . . . . . . . . . . . . .pounds (Avoir.) per gallon (Imp.—liquid)

## PARTS PER MILLION (cont'd): =

| | |
|---|---|
| 0.000160355 | ounces (Avoir.) per gallon (Imp.—liquid) |
| 0.00454597 | grams per gallon (Imperial—liquid) |
| 0.0701552 | grains per gallon (Imperial—liquid) |
| 0.00000000110231 | tons (net) per liter |
| 0.000001 | kilograms per liter |
| 0.00000220462 | pounds (Avoir.) per liter |
| 0.0000352739 | ounces (Avoir.) per liter |
| 0.001 | grams per liter |
| 0.0154324 | grains per liter |
| 0.0000000000180633 | tons (net) per cubic inch |
| 0.0000000163867 | kilograms per cubic inch |
| 0.0000000361264 | pounds (Avoir.) per cubic inch |
| 0.000000578023 | ounces (Avoir.) per cubic inch |
| 0.0000163867 | grams per cubic inch |
| 0.000252886 | grains per cubic inch |
| 8.345 | pounds per million gallons |

## POISE: =

| | |
|---|---|
| 1 | gram per centimeter per second |

## PENNYWEIGHT: =

| | |
|---|---|
| 0.00000153061 | tons (long) |
| 0.00000155517 | tons (metric) |
| 0.00000171429 | tons (net) |
| 0.00155517 | kilograms |
| 0.0041667 | pounds (Troy) |
| 0.00342857 | pounds (Avoir.) |
| 0.0155517 | hektograms |
| 0.155517 | dekagrams |
| 0.05 | ounces (Troy) |
| 0.0548571 | ounces (Avoir.) |
| 1.55517 | grams |
| 15.5517 | decigrams |
| 155.517 | centigrams |
| 1,555.17 | milligrams |
| 24 | grains |
| 0.4 | drams (Troy) |
| 0.877714 | drams (Avoir.) |
| 1 | pennyweights |
| 1.2 | scruples |
| 7.776 | carats (metric) |

POUND (TROY): =

|  |  |
|---|---|
| 0.000367347 | tons (long) |
| 0.000373242 | tons (metric) |
| 0.000411429 | tons (net) |
| 0.373242 | kilograms |
| 1 | pounds (Troy) |
| 0.822857 | pounds (Avoir.) |
| 3.732418 | hektograms |
| 37.324176 | dekagrams |
| 12 | ounces (Troy) |
| 13.165714 | ounces (Avoir.) |
| 373.241762 | grams |
| 3,732.417621 | decigrams |
| 37,324.176213 | centigrams |
| 373,242 | milligrams |
| 5,760 | grains |
| 96 | drams (Troy) |
| 210.651425 | drams (Avoir.) |
| 240 | pennyweights |
| 288 | scruples |
| 1,866.239964 | carats (metric) |

POUND (AVOIRDUPOIS): =

|  |  |
|---|---|
| 0.000446429 | tons (long) |
| 0.000453593 | tons (metric) |
| 0.0005 | tons (net) |
| 0.453592 | kilograms |
| 1.215278 | pounds (Troy) |
| 1 | pounds (Avoir.) |
| 4.535924 | hektograms |
| 45.359243 | dekagrams |
| 14.5833 | ounces (Troy) |
| 16 | ounces (Avoir.) |
| 453.592428 | grams |
| 4,535.92428 | decigrams |
| 45,359.2428 | centigrams |
| 453,592 | milligrams |
| 7,000 | grains |
| 116.666675 | drams (Troy) |
| 256 | drams (Avoir.) |
| 291.6667 | pennyweights |

POUND (AVOIRDUPOIS) (cont'd): =

| | |
|---|---|
| 350.1 | scruples |
| 2,268 | carats (metric) |

POUND OF CARBON OXIDIZED WITH 100% EFFICIENCY: =

| | |
|---|---|
| 365,245,567 | foot poundals |
| 4,382,946,806 | inch poundals |
| 11,352,418 | foot pounds |
| 136,229,016 | inch pounds |
| 3.676937 | ton calories |
| 3,676.937455 | kilogram calories |
| 8,106.271602 | pound calories |
| 36,769.374545 | hektogram calories |
| 367,694 | dekagram calories |
| 129,700 | ounce calories |
| 3,676,937 | gram calories (mean) |
| 36,769,375 | decigram calories |
| 367,693,745 | centigram calories |
| 3,676,937,455 | milligram calories |
| 0.178109 | kilowatt days |
| 4.274614 | kilowatt hours |
| 256.477745 | kilowatt minutes |
| 15,388.634345 | kilowatt seconds |
| 178.109039 | watt days |
| 4,274.613901 | watt hours |
| 256,477 | watt minutes |
| 15,388,624 | watt seconds |
| 1,569.531035 | ton meters |
| 1,569,513 | kilogram meters |
| 3,460,223 | pounds meters |
| 15,695,310 | hektogram meters |
| 156,953,104 | dekagram meters |
| 55,363,570 | ounce meters |
| 1,569,531,035 | gram meters |
| 15,695,310,350 | decigram meters |
| 156,953,103,500 | centigram meters |
| $1,569,531,035 \times 10^3$ | milligram meters |
| 15.695310 | ton hektometers |
| 15,695.31035 | kilogram hektometers |
| 34,602.288580 | pound hektometers |
| 156,953 | hektogram hektometers |
| 1,569,531 | dekagram hektometers |

## POUND OF CARBON OXIDIZED WITH 100% EFFICIENCY (cont'd): =

553,636 . . . . . . . . . . . . . . . . . . . . . . . . . . ounce hektometers
15,695,310 . . . . . . . . . . . . . . . . . . . . . . . . . gram hektometers
156,953,104 . . . . . . . . . . . . . . . . . . . . . . . .decigram hektometers
1,569,531,035 . . . . . . . . . . . . . . . . . . . . . . centigram hektometers
15,695,310,350 . . . . . . . . . . . . . . . . . . . . . milligram hektometers
156.907535 . . . . . . . . . . . . . . . . . . . . . . . . . .ton dekameters
156,907 . . . . . . . . . . . . . . . . . . . . . . . . . kilogram dekameters
346,022 . . . . . . . . . . . . . . . . . . . . . . . . . . .pound dekameters
1,569,075 . . . . . . . . . . . . . . . . . . . . . . . . hektogram dekameters
15,690,754 . . . . . . . . . . . . . . . . . . . . . . . . dekagram dekameters
5,536,358 . . . . . . . . . . . . . . . . . . . . . . . . . . ounce dekameters
156,907,535 . . . . . . . . . . . . . . . . . . . . . . . . . .gram dekameters
1,569,075,350 . . . . . . . . . . . . . . . . . . . . . . decigram dekameters
15,690,753,500 . . . . . . . . . . . . . . . . . . . . . .centigram dekameters
$156,907,535 \times 10^3$ . . . . . . . . . . . . . . . . . milligram dekameters
5,149.255690 . . . . . . . . . . . . . . . . . . . . . . . . . . . . ton feet
5,149,256 . . . . . . . . . . . . . . . . . . . . . . . . . . . . kilogram feet
11,352,449 . . . . . . . . . . . . . . . . . . . . . . . . . . . . pound feet
51,492,557 . . . . . . . . . . . . . . . . . . . . . . . . . . .hektogram feet
514,925,569 . . . . . . . . . . . . . . . . . . . . . . . . . . dekagram feet
181,639,190 . . . . . . . . . . . . . . . . . . . . . . . . . . . . ounce feet
5,149,255,690 . . . . . . . . . . . . . . . . . . . . . . . . . . . . gram feet
51,492,556,895 . . . . . . . . . . . . . . . . . . . . . . . . .decigram feet
514,925,568,950 . . . . . . . . . . . . . . . . . . . . . . . . centigram feet
$51,492,556,895 \times 10^2$ . . . . . . . . . . . . . . . . . . . milligram feet
61,792,556845 . . . . . . . . . . . . . . . . . . . . . . . . . . . .ton inches
61,792,557 . . . . . . . . . . . . . . . . . . . . . . . . . . kilogram inches
136,229,386 . . . . . . . . . . . . . . . . . . . . . . . . . . . pound inches
617,925,568 . . . . . . . . . . . . . . . . . . . . . . . . . hektogram inches
6,179,255,685 . . . . . . . . . . . . . . . . . . . . . . . . dekagram inches
2,179,670,177 . . . . . . . . . . . . . . . . . . . . . . . . . . ounce inches
61,792,556,845 . . . . . . . . . . . . . . . . . . . . . . . . . . .gram inches
617,925,568,450 . . . . . . . . . . . . . . . . . . . . . . . . decigram inches
$61,792,556,845 \times 10^2$ . . . . . . . . . . . . . . . . . . centigram inches
$61,792,556,845 \times 10^3$ . . . . . . . . . . . . . . . . . . milligram inches
15,695.31035 . . . . . . . . . . . . . . . . . . . . . . . . . . ton decimeters
15,695,310 . . . . . . . . . . . . . . . . . . . . . . . . . kilogram decimeters
34,602,231 . . . . . . . . . . . . . . . . . . . . . . . . . . pound decimeters
156,953,104 . . . . . . . . . . . . . . . . . . . . . . . .hektogram decimeters
1,569,531,035 . . . . . . . . . . . . . . . . . . . . . . dekagram decimeters
553,635,699 . . . . . . . . . . . . . . . . . . . . . . . . . ounce decimeters
15,695,310,350 . . . . . . . . . . . . . . . . . . . . . . . . gram decimeters
156,953,103,500 . . . . . . . . . . . . . . . . . . . . . .decigram decimeters
$1,569,531,035 \times 10^3$ . . . . . . . . . . . . . . . . centigram decimeters

## POUND OF CARBON OXIDIZED WITH 100% EFFICIENCY (cont'd): =

$1,569,531,035 \times 10^4$ . . . . . . . . . . . . . . . . . . . . . . milligram decimeters
156,953 . . . . . . . . . . . . . . . . . . . . . . . . . . . . . . . ton centimeters
156,953,104 . . . . . . . . . . . . . . . . . . . . . . . . . kilogram centimeters
346,022,312 . . . . . . . . . . . . . . . . . . . . . . . . .pound centimeters
1,569,531,035 . . . . . . . . . . . . . . . . . . . . . . hektogram centimeters
15,695,310,350 . . . . . . . . . . . . . . . . . . . . .dekagram centimeters
5,536,356,992 . . . . . . . . . . . . . . . . . . . . . . . . ounce centimeters
156,953,103,500 . . . . . . . . . . . . . . . . . . . . . .gram centimeters
$1,569,531,035 \times 10^3$ . . . . . . . . . . . . . . . . . . decigram centimeters
$1,569,531,035 \times 10^4$ . . . . . . . . . . . . . . . . . .centigram centimeters
$1,569,531,035 \times 10^5$ . . . . . . . . . . . . . . . . . . milligram centimeters
1,569,531 . . . . . . . . . . . . . . . . . . . . . . . . . . . . . .ton millimeters
1,569,531,035 . . . . . . . . . . . . . . . . . . . . . . . .kilogram millimeters
3,460,223,121 . . . . . . . . . . . . . . . . . . . . . . . . . pound millimeters
15,695,310,350 . . . . . . . . . . . . . . . . . . . . . hektogram millimeters
156,953,103,500 . . . . . . . . . . . . . . . . . . . . . dekagram millimeters
55,363,569,923 . . . . . . . . . . . . . . . . . . . . . . . . ounce millimeters
$1,569,531,035 \times 10^3$ . . . . . . . . . . . . . . . . . . . . . . gram millimeters
$1,569,531,035 \times 10^4$ . . . . . . . . . . . . . . . . . . . decigram millimeters
$1,569,531,035 \times 10^5$ . . . . . . . . . . . . . . . . . . centigram millimeters
$1,569,531,035 \times 10^6$ . . . . . . . . . . . . . . . . . . milligram millimeters
151.895 . . . . . . . . . . . . . . . . . . . . . . . . . . . kiloliter-atmospheres
1,518.95 . . . . . . . . . . . . . . . . . . . . . . . . . hektoliter-atmospheres
15,189.5 . . . . . . . . . . . . . . . . . . . . . . . . . . dekaliter-atmospheres
151,895 . . . . . . . . . . . . . . . . . . . . . . . . . . . . . .liter-atmospheres
1,518,950 . . . . . . . . . . . . . . . . . . . . . . . . . . deciliter-atmospheres
15,189,500 . . . . . . . . . . . . . . . . . . . . . . . . .centiliter-atmospheres
151,895,000 . . . . . . . . . . . . . . . . . . . . . . . . . milliliter-atmospheres
0.000000151895 . . . . . . . . . . . . . . . . . . . .cubic kilometer-atmospheres
0.000151895 . . . . . . . . . . . . . . . . . . . . . .cubic hektometer atmospheres
0.151895 . . . . . . . . . . . . . . . . . . . . . . . . cubic dekameter-atmospheres
151.895 . . . . . . . . . . . . . . . . . . . . . . . . . . . cubic meter-atmospheres
5,364.779505 . . . . . . . . . . . . . . . . . . . . . . . . cubic foot-atmospheres
9,269,392 . . . . . . . . . . . . . . . . . . . . . . . . . . . cubic inch-atmospheres
151,895 . . . . . . . . . . . . . . . . . . . . . . . . .cubic decimeter-atmospheres
151,895,000 . . . . . . . . . . . . . . . . . . . . . . cubic centimeter-atmospheres
$151,895 \times 10^6$ . . . . . . . . . . . . . . . . . . . . cubic millimeter-atmospheres
15,391,217 . . . . . . . . . . . . . . . . . . . . . . . . . . . . . .joules (absolute)
5.813022 . . . . . . . . . . . . . . . . . . . . . . . . . . . . Cheval-Vapeur hours
0.238896 . . . . . . . . . . . . . . . . . . . . . . . . . . . . . . . horsepower days
5.733277 . . . . . . . . . . . . . . . . . . . . . . . . . . . . . . horsepower hours
344.0102771 . . . . . . . . . . . . . . . . . . . . . . . . . . .horsepower minutes
20,640.555865 . . . . . . . . . . . . . . . . . . . . . . . . . horsepower seconds
1 . . . . . . . . . . . . . . . . . pounds of carbon oxidized with 100% efficiency

POUND OF CARBON OXIDIZED WITH 100% EFFICIENCY (cont'd): =

15.0482377 . . . . . . . . . pounds of water evaporated from and at 212° F.
6,618,631 . . . . . . . . . . . . . . . . . . . . . . . . . . BTU (mean) per pound
14,592.55265 . . . . . . . . . . . . . . . . . . . . . . . . . . . . . . . BTU (mean)
$153{,}918{,}415{,}320 \times 10^3$ . . . . . . . . . . . . . . . . . . . . . . . . . . . . . . ergs

POUND (PRESSURE) PER SQUARE INCH: =

0.0070307 . . . . . . . . . . . . . . . . . . . . . . . .hektometers of water @ 60° F.
0.070307 . . . . . . . . . . . . . . . . . . . . . . . dekameters of water @ 60° F.
0.70307 . . . . . . . . . . . . . . . . . . . . . . . . . . meters of water @ 60° F.
2.306787 . . . . . . . . . . . . . . . . . . . . . . . . . . .feet of water @ 60° F.
27.681473 . . . . . . . . . . . . . . . . . . . . . . . . . inches of water @ 60° F.
7.0307 . . . . . . . . . . . . . . . . . . . . . . . . .decimeters of water @ 60° F.
70,307 . . . . . . . . . . . . . . . . . . . . . . . . centimeters of water @ 60° F.
703.07 . . . . . . . . . . . . . . . . . . . . . . . . millimeters of water @ 60° F.
0.00051715 . . . . . . . . . . . . . . . . . . . . hektometers of mercury @ 32° F.
0.0051715 . . . . . . . . . . . . . . . . . . . . . .dekameters of mercury @ 32° F.
0.051715 . . . . . . . . . . . . . . . . . . . . . . . . . meters of mercury @ 32° F.
0.169667 . . . . . . . . . . . . . . . . . . . . . . . . . . . feet of mercury @ 32° F.
2.0360 . . . . . . . . . . . . . . . . . . . . . . . . . . .inches of mercury @ 32° F.
0.51715 . . . . . . . . . . . . . . . . . . . . . . . . . decimeters of mercury @ 32° F.
5.1715 . . . . . . . . . . . . . . . . . . . . . . . . . centimeters of mercuy @ 32° F.
51.715 . . . . . . . . . . . . . . . . . . . . . . . . .millimeters of mercury @ 32° F.
7,750.0696898 . . . . . . . . . . . . . . . . . . . . . . . .tons per square hektometer
77.500697 . . . . . . . . . . . . . . . . . . . . . . . . . tons per square dekameter
0.775007 . . . . . . . . . . . . . . . . . . . . . . . . . . . . . tons per square meter
0.0720023 . . . . . . . . . . . . . . . . . . . . . . . . . . . . . tons per square foot
0.0005 . . . . . . . . . . . . . . . . . . . . . . . . . . . . . . . tons per square inch
0.00775007 . . . . . . . . . . . . . . . . . . . . . . . . . .tons per square decimeter
0.0000775007 . . . . . . . . . . . . . . . . . . . . . . . tons per square centimeter
0.000000775007 . . . . . . . . . . . . . . . . . . . . . tons per square millimeter
7,030,700 . . . . . . . . . . . . . . . . . . . . . . kilograms per square hektometer
70,307 . . . . . . . . . . . . . . . . . . . . . . . . kilograms per square hektometer
703.07 . . . . . . . . . . . . . . . . . . . . . . . . . . . kilograms per square meter
65.317203 . . . . . . . . . . . . . . . . . . . . . . . . . . .kilograms per square foot
0.453592 . . . . . . . . . . . . . . . . . . . . . . . . . . .kilograms per square inch
7.0307 . . . . . . . . . . . . . . . . . . . . . . . . . kilograms per square decimeter
0.070307 . . . . . . . . . . . . . . . . . . . . . . . kilograms per square centimeter
0.00070307 . . . . . . . . . . . . . . . . . . . . . . kilograms per square millimeter
15,500,130 . . . . . . . . . . . . . . . . . . . . . . . . pounds per square hektometer
155,001 . . . . . . . . . . . . . . . . . . . . . . . . . . pounds per square dekameter
1,550.0130 . . . . . . . . . . . . . . . . . . . . . . . . . . . . pounds per square meter
144 . . . . . . . . . . . . . . . . . . . . . . . . . . . . . . . . . . pounds per square foot

POUND (PRESSURE) PER SQUARE INCH (cont'd): =

|  |  |
|---|---|
| 1 | pounds per square inch |
| 15.500130 | pounds per square decimeter |
| 0.155001 | pounds per square centimeter |
| 0.00155001 | pounds per square millimeter |
| 70,307,000 | hektograms per square hektometer |
| 703,070 | hektograms per square dekameter |
| 7,030.70 | hektograms per square meter |
| 653.172168 | hektograms per square foot |
| 4.535933 | hektograms per square inch |
| 70.3070 | hektograms per square decimeter |
| 0.70307 | hektograms per square centimeter |
| 0.0070307 | hektograms per square millimeter |
| 703,070,000 | dekagrams per square hektometer |
| 7,030,700 | dekagrams per square dekameter |
| 70,307 | dekagrams per square meter |
| 6,531.721544 | dekagrams per square foot |
| 45.359332 | dekagrams per square inch |
| 7,030.7 | dekagrams per square decimeter |
| 70.307 | dekagrams per square centimeter |
| 0.70307 | dekagrams per square millimeter |
| 248,002,217 | ounces per square hektometer |
| 2,480,022 | ounces per square dekameter |
| 24,800.22 | ounces per square meter |
| 2,303.985941 | ounces per square foot |
| 16 | ounces per square inch |
| 248.0022 | ounces per square decimeter |
| 2.480022 | ounces per square centimeter |
| 0.0248002 | ounces per square millimeter |
| 7,030,700,000 | grams per square hektometer |
| 70,307,000 | grams per square dekameter |
| 703,070 | grams per square meter |
| 65,317.215135 | grams per square foot |
| 453.593927 | grams per square inch |
| 7,030.7 | grams per square decimeter |
| 70.307 | grams per square centimeter |
| 0.70307 | grams per square millimeter |
| 70,307,000,000 | decigrams per square hektometer |
| 703,070,000 | decigrams per square dekameter |
| 7,030,700 | decigrams per square meter |
| 653,172 | decigrams per square foot |
| 4,535.939265 | decigrams per square inch |
| 70,307 | decigrams per square decimeter |
| 703.07 | decigrams per square centimeter |
| 7.0307 | decigrams per square millimeter |

POUND (PRESSURE) PER SQUARE INCH (cont'd): =

| | |
|---|---|
| 703,070,000,000 | centigrams per square hektometer |
| 7,030,700,000 | centigrams per square dekameter |
| 70,307,000 | centigrams per square meter |
| 6,531,720 | centigrams per square foot |
| 45,359.392595 | centigrams per square inch |
| 703,070 | centigrams per square decimeter |
| 7,030.70 | centigrams per square centimeter |
| 70.3070 | centigrams per square millimeter |
| $70{,}307 \times 10^8$ | milligrams per square hektometer |
| $70{,}307 \times 10^6$ | milligrams per square dekameter |
| $70{,}307 \times 10^4$ | milligrams per square meter |
| 65,317,200 | milligrams per square foot |
| 453,594 | milligrams per square inch |
| 70,307,000 | milligrams per square decimeter |
| 703,070 | milligrams per square centimeter |
| 7,030.70 | milligrams per square millimeter |
| $68{,}947 \times 10^8$ | dynes per square hektometer |
| 68,947,000,000 | dynes per square dekameter |
| 689,470,000 | dynes per square meter |
| 64,053,184 | dynes per square foot |
| 448,814 | dynes per square inch |
| 6,894,700 | dynes per square decimeter |
| 68,947 | dynes per square centimeter |
| 689.47 | dynes per square millimeter |
| 0.068947 | bars |
| 0.068046 | atmospheres |

POUND (WEIGHT) PER SQUARE FOOT: =

| | |
|---|---|
| 0.0000488243 | hektometers of water @ 60° F. |
| 0.000488243 | dekameters of water @ 60° F. |
| 0.00488243 | meters of water @ 60° F. |
| 0.0160194 | feet of water @ 60° F. |
| 0.192232 | inches of water @ 60° F. |
| 0.0488243 | decimeters of water @ 60° F. |
| 0.488243 | centimeters of water @ 60° F. |
| 4.882431 | millimeters of water @ 60° F. |
| 0.00000359132 | hektometers of mercury @ 32° F. |
| 0.0000359132 | dekameters of mercury @ 32° F. |
| 0.000359132 | meters of mercury @ 32° F. |
| 0.00117824 | feet of mercury @ 32° F. |
| 0.0141389 | inches of mercury @ 32° F. |
| 0.00359132 | decimeters of mercury @ 32° F. |

POUND (WEIGHT) PER SQUARE FOOT (cont'd): =

| | |
|---|---|
| 0.0359132 | centimeters of mercury @ 32° F. |
| 0.359132 | millimeters of mercury @ 32° F. |
| 53.819930 | tons per square hektometer |
| 0.538199 | tons per square dekameter |
| 0.00538199 | tons per square meter |
| 0.0005000016 | tons per square foot |
| 0.00000347222 | tons per square inch |
| 0.0000538199 | tons per square decimeter |
| 0.000000538199 | tons per square centimeter |
| 0.00000000538199 | tons per square millimeter |
| 48,824.305552 | kilograms per square hektometer |
| 488.243056 | kilograms per square dekameter |
| 4.882431 | kilograms per square meter |
| 0.453592 | kilograms per square foot |
| 0.00314994 | kilograms per square inch |
| 0.0488243 | kilograms per square decimeter |
| 0.000488243 | kilograms per square centimeter |
| 0.00000488243 | kilograms per square millimeter |
| 107,640 | pounds per square hektometer |
| 1,076.397917 | pounds per square dekameter |
| 10.763979 | pounds per square meter |
| 1 | pounds per square foot |
| 0.00694444 | pounds per square inch |
| 0.107640 | pounds per square decimeter |
| 0.00107640 | pounds per square centimeter |
| 0.0000107640 | pounds per square millimeter |
| 488,243 | hektograms per square hektometer |
| 4,882.430555 | hektograms per square dekameter |
| 48.824306 | hektograms per square meter |
| 4.535918 | hektograms per square foot |
| 0.0314994 | hektograms per square inch |
| 0.488243 | hektograms per square decimeter |
| 0.00488243 | hektograms per square centimeter |
| 0.0000488243 | hektograms per square millimeter |
| 4,882,431 | dekagrams per square hektometer |
| 48,824.305552 | dekagrams per square dekameter |
| 488.243056 | dekagrams per square meter |
| 45.359178 | dekagrams per square foot |
| 0.314994 | dekagrams per square inch |
| 4.882431 | dekagrams per square decimeter |
| 0.0488243 | dekagrams per square centimeter |
| 0.000488243 | dekagrams per square millimeter |
| 1,722,238 | ounces per square hektometer |
| 17,222.376179 | ounces per square dekameter |

POUND WEIGHT PER SQUARE FOOT (cont'd): =

| | |
|---|---|
| 172.223762 | ounces per square meter |
| 16 | ounces per square foot |
| 0.111111 | ounces per square inch |
| 1.722238 | ounces per square decimeter |
| 0.0172224 | ounces per square centimeter |
| 0.000172224 | ounces per square millimeter |
| 48,824,306 | grams per square hektometer |
| 488,243 | grams per square dekameter |
| 4,882.430555 | grams per square meter |
| 453.591783 | grams per square foot |
| 3.149953 | grams per square inch |
| 48.824306 | grams per square decimeter |
| 0.488243 | grams per square centimeter |
| 0.00488243 | grams per square millimeter |
| 488,243,056 | decigrams per square hektometer |
| 4,882,431 | decigrams per square dekameter |
| 48,824.305552 | decigrams per square meter |
| 4,535.917833 | decigrams per square foot |
| 3.149953 | decigrams per square inch |
| 488.243056 | decigrams per square decimeter |
| 4.882431 | decigrams per square centimeter |
| 0.0488243 | decigrams per square millimeter |
| 4,882,430,555 | centigrams per square hektometer |
| 48,824,306 | centigrams per square dekameter |
| 488,243 | centigrams per square meter |
| 45,359.178330 | centigrams per square foot |
| 31.499535 | centigrams per square inch |
| 4,882.430555 | centigrams per square decimeter |
| 48.824306 | centigrams per square centimeter |
| 0.488243 | centigrams per square millimeter |
| 48,824,305,552 | milligrams per square hektometer |
| 488,243,056 | milligrams per square dekameter |
| 4,882.431 | milligrams per square meter |
| 453,592 | milligrams per square foot |
| 314.995347 | milligrams per square inch |
| 48,824.305552 | milligrams per square decimeter |
| 488.243056 | milligrams per square centimeter |
| 4.882431 | milligrams per square millimeter |
| 47,879,860,000 | dynes per square hektometer |
| 478,798,600 | dynes per square dekameter |
| 4,787,986 | dynes per square meter |
| 444,814 | dynes per square foot |
| 3,116.763889 | dynes per square inch |
| 47,879.861108 | dynes per square decimeter |

POUND (WEIGHT) PER SQUARE FOOT (cont'd): =

| | |
|---|---|
| 478.798611 | dynes per square centimeter |
| 4.787986 | dynes per square millimeter |
| 0.00047880 | bars |
| 0.00047254 | atmospheres |

POUND PER INCH: =

| | |
|---|---|
| 19.685057 | tons (net) per kilometer |
| 31.680146 | tons (net) per mile |
| 0.0196851 | tons (net) per meter |
| 0.018 | tons (net) per yard |
| 0.006 | tons (net) per foot |
| 0.0005 | tons (net) per inch |
| 0.000196851 | tons (net) per centimeter |
| 0.0000196851 | tons (net) per millimeter |
| 17,857.985915 | kilograms per kilometer |
| 28,739.749194 | kilograms per mile |
| 17.857985 | kilograms per meter |
| 16.329328 | kilograms per yard |
| 5.443109 | kilograms per foot |
| 0.453592 | kilograms per inch |
| 0.178579 | kilograms per centimeter |
| 0.0178579 | kilograms per millimeter |
| 39,370.11300 | pounds (Avoir.) per kilometer |
| 63,360.291357 | pounds (Avoir.) per mile |
| 39.370113 | pounds (Avoir.) per meter |
| 36 | pounds (Avoir.) per yard |
| 12 | pounds (Avoir.) per foot |
| 1 | pounds (Avoir.) per inch |
| 0.393701 | pounds (Avoir.) per centimeter |
| 0.0393701 | pounds (Avoir.) per millimeter |
| 629,921.80800 | ounces (Avoir.) per kilometer |
| 1,013,764 | ounces (Avoir.) per mile |
| 629.921808 | ounces (Avoir.) per meter |
| 576 | ounces (Avoir.) per yard |
| 192 | ounces (Avoir.) per foot |
| 16 | ounces (Avoir.) per inch |
| 6.299218 | ounces (Avoir.) per centimeter |
| 0.629921 | ounces (Avoir.) per millimeter |
| 17,857,985 | grams per kilometer |
| 28,739,794 | grams per mile |
| 17,857.9851 | grams per meter |
| 16,329.327396 | grams per yard |

POUND PER INCH (cont'd): =

5,443.109132 .......... grams per foot
453.5924277 .......... grams per inch
178.579851 .......... grams per centimeter
17.857985 .......... grams per millimeter
275,590,791 .......... grains per kilometer
443,522,039 .......... grains per mile
275,590.791 .......... grains per meter
252,000 .......... grains per yard
84,000 .......... grains per foot
7,000 .......... grains per inch
2,755.90791 .......... grains per centimeter
275.590791 .......... grains per millimeter

POUND (AVOIR.) PER FOOT: =

1.640239 .......... tons (net) per kilometer
2.639719 .......... tons (net) per mile
0.00164024 .......... tons (net) per meter
0.00149984 .......... tons (net) per yard
0.000499946 .......... tons (net) per foot
0.0000416622 .......... tons (net) per inch
0.0000164024 .......... tons (net) per centimeter
0.00000164024 .......... tons (net) per millimeter
1,488.161203 .......... kilograms per kilometer
2,394.71280 .......... kilograms per mile
1.488161 .......... kilograms per meter
1.360777 .......... kilograms per yard
0.453592 .......... kilograms per foot
0.0377994 .......... kilograms per inch
0.0148816 .......... kilograms per centimeter
0.00148816 .......... kilograms per millimeter
3,280.8 .......... pounds (Avoir.) per kilometer
5,280 .......... pounds (Avoir.) per mile
3.280833 .......... pounds (Avoir.) per meter
3 .......... pounds (Avoir.) per yard
1 .......... pounds (Avoir.) per foot
0.0833333 .......... pounds (Avoir.) per inch
0.0328083 .......... pounds (Avoir.) per centimeter
0.00328083 .......... pounds (Avoir.) per millimeter
52,493.328 .......... ounces (Avoir.) per kilometer
84,480 .......... ounces (Avoir.) per mile
52.493328 .......... ounces (Avoir.) per meter
48 .......... ounces (Avoir.) per yard

**POUND (AVOIR.) PER FOOT (cont'd): =**

| | |
|---|---|
| 16 | ounces (Avoir.) per foot |
| 1.333333 | ounces (Avoir.) per inch |
| 0.524933 | ounces (Avoir.) per centimeter |
| 0.0524933 | ounces (Avoir.) per millimeter |
| 1,488,161 | grams per kilometer |
| 2,394,713 | grams per mile |
| 1,488.161 | grams per meter |
| 1,360.777283 | grams per yard |
| 453.5924277 | grams per foot |
| 37.799369 | grams per inch |
| 14.881612 | grams per centimeter |
| 1.488161 | grams per millimeter |
| 22,965,899 | grains per kilometer |
| 36,960,000 | grains per mile |
| 22,965.899250 | grains per meter |
| 21,000 | grains per yard |
| 7,000 | grains per foot |
| 583.333333 | grains per inch |
| 229.658993 | grains per centimeter |
| 22.965899 | grains per millimeter |

**POUND (WEIGHT) PER CUBIC FOOT: =**

| | |
|---|---|
| 17,657.261726 | tons (net) per cubic hektometer |
| 17.657262 | tons (net) per cubic dekameter |
| 0.01765731 | tons (net) per cubic meter |
| 0.0005 | tons (net) per cubic foot |
| 0.000000289352 | tons (net) per cubic inch |
| 0.0000176573 | tons (net) per cubic decimeter |
| 0.0000000176573 | tons (net) per cubic centimeter |
| 0.0000000000176573 | tons (net) per cubic millimeter |
| 16,018,400 | kilograms per cubic hektometer |
| 16,018.4 | kilograms per cubic dekameter |
| 16,0184 | kilograms per cubic meter |
| 0.453593 | kilograms per cubic foot |
| 0.000262496 | kilograms per cubic inch |
| 0.0160184 | kilograms per cubic decimeter |
| 0.0000160184 | kilograms per cubic centimeter |
| 0.0000000160184 | kilograms per cubic millimeter |
| 35,314,445 | pounds per cubic hektometer |
| 35,314.445 | pounds per cubic dekameter |
| 35.314445 | pounds per cubic meter |
| 1 | pounds per cubic foot |

POUND (WEIGHT) PER CUBIC FOOT (cont'd): =

0.000578704 . . . . . . . . . . . . . . . . . . . . . . . . .pounds per cubic inch
0.0353144 . . . . . . . . . . . . . . . . . . . . . . . pounds per cubic decimeter
0.0000353144 . . . . . . . . . . . . . . . . . . . . . pounds per cubic centimeter
0.0000000353144 . . . . . . . . . . . . . . . . . . pounds per cubic millimeter
160,184,000 . . . . . . . . . . . . . . . . . . . . hektograms per cubic hektometer
160,184 . . . . . . . . . . . . . . . . . . . . . . . .hektograms per cubic dekameter
160.184 . . . . . . . . . . . . . . . . . . . . . . . . . hektograms per cubic meter
4.535934 . . . . . . . . . . . . . . . . . . . . . . . . . hektograms per cubic foot
0.00262496 . . . . . . . . . . . . . . . . . . . . . . . hektograms per cubic inch
0.160184 . . . . . . . . . . . . . . . . . . . . . . . . hektograms per cubic decimeter
0.000160184 . . . . . . . . . . . . . . . . . . . . . hektograms per cubic centimeter
0.000000160184 . . . . . . . . . . . . . . . . . .hektograms per cubic millimeter
1,601,840,000 . . . . . . . . . . . . . . . . . . . .dekagrams per cubic hektometer
1,601,840 . . . . . . . . . . . . . . . . . . . . . . . dekagrams per cubic dekameter
1,601.840 . . . . . . . . . . . . . . . . . . . . . . . . . dekagrams per cubic meter
45.359342 . . . . . . . . . . . . . . . . . . . . . . . . .dekagrams per cubic foot
0.0262496 . . . . . . . . . . . . . . . . . . . . . . . . .dekagrams per cubic inch
1.60184 . . . . . . . . . . . . . . . . . . . . . . . . .dekagrams per cubic decimeter
0.00160184 . . . . . . . . . . . . . . . . . . . . . . dekagrams per cubic centimeter
0.00000160184 . . . . . . . . . . . . . . . . . . . dekagrams per cubic millimeter
565,031,120 . . . . . . . . . . . . . . . . . . . . . . . .ounces per cubic hektometer
565,031 . . . . . . . . . . . . . . . . . . . . . . . . . . ounces per cubic dekameter
565.031120 . . . . . . . . . . . . . . . . . . . . . . . . . . ounces per cubic meter
16 . . . . . . . . . . . . . . . . . . . . . . . . . . . . . . . . ounces per cubic foot
0.00925926 . . . . . . . . . . . . . . . . . . . . . . . . . . ounces per cubic inch
0.565031 . . . . . . . . . . . . . . . . . . . . . . . . . .ounces per cubic decimeter
0.000565031 . . . . . . . . . . . . . . . . . . . . . . . ounces per cubic centimeter
0.000000565031 . . . . . . . . . . . . . . . . . . . . ounces per cubic millimeter
16,018,400,000 . . . . . . . . . . . . . . . . . . . . . grams per cubic hektometer
16,018,400 . . . . . . . . . . . . . . . . . . . . . . . . . grams per cubic dekameter
16,018.4 . . . . . . . . . . . . . . . . . . . . . . . . . . . . .grams per cubic meter
453.593422 . . . . . . . . . . . . . . . . . . . . . . . . . . .grams per cubic foot
0.262496 . . . . . . . . . . . . . . . . . . . . . . . . . . . .grams per cubic inch
16.0184 . . . . . . . . . . . . . . . . . . . . . . . . . . grams per cubic decimeter
0.0160184 . . . . . . . . . . . . . . . . . . . . . . . . . grams per cubic centimeter
0.0000160184 . . . . . . . . . . . . . . . . . . . . . . grams per cubic millimeter
160,184,000,000 . . . . . . . . . . . . . . . . . . decigrams per cubic hektometer
160,184,000 . . . . . . . . . . . . . . . . . . . . . . .decigrams per cubic dekameter
160,184 . . . . . . . . . . . . . . . . . . . . . . . . . . . decigrams per cubic meter
4,535.934220 . . . . . . . . . . . . . . . . . . . . . . . . decigrams per cubic foot
2.624962 . . . . . . . . . . . . . . . . . . . . . . . . . . . decigrams per cubic inch
160.184 . . . . . . . . . . . . . . . . . . . . . . . . . decigrams per cubic decimeter
0.160184 . . . . . . . . . . . . . . . . . . . . . . . . decigrams per cubic centimeter
0.000160184 . . . . . . . . . . . . . . . . . . . . . .decigrams per cubic millimeter

POUND (WEIGHT) PER CUBIC FOOT (cont'd): =

| | |
|---|---|
| $160,184 \times 10^7$ | centigrams per cubic hektometer |
| 1,601,840,000 | centigrams per cubic dekameter |
| 1,601,840 | centigrams per cubic meter |
| 45,359.342205 | centigrams per cubic foot |
| 26.249619 | centigrams per cubic inch |
| 1,601.840 | centigrams per cubic decimeter |
| 1.60184 | centigrams per cubic centimeter |
| 0.00160184 | centigrams per cubic millimeter |
| $160,184 \times 10^8$ | milligrams per cubic hektometer |
| 16,018,400,000 | milligrams per cubic dekameter |
| 16,018,400 | milligrams per cubic meter |
| 453,593 | milligrams per cubic foot |
| 262.496193 | milligrams per cubic inch |
| 16,018.4 | milligrams per cubic decimeter |
| 16.0184 | milligrams per cubic centimeter |
| 0.0160184 | milligrams per cubic millimeter |
| $157,085,886 \times 10^9$ | dynes per cubic hektometer |
| $157,086 \times 10^6$ | dynes per cubic dekameter |
| 157,086,000 | dynes per cubic meter |
| 444,820 | dynes per cubic foot |
| 257.418981 | dynes per cubic inch |
| 157,086 | dynes per cubic decimeter |
| 157.0858859 | dynes per cubic centimeter |
| 0.157086 | dynes per cubic millimeter |

POUND PER CUBIC INCH: =

| | |
|---|---|
| 30,511,748 | tons (net) per cubic hektometer |
| 30,511.748042 | tons (net) per cubic dekameter |
| 30.511748 | tons (net) per cubic meter |
| 0.864001 | tons (net) per cubic foot |
| 0.0005 | tons (net) per cubic inch |
| 0.0305117 | tons (net) per cubic decimeter |
| 0.0000305117 | tons (net) per cubic centimeter |
| 0.0000000305117 | tons (net) per cubic millimeter |
| 27,679,795,200 | kilograms per cubic hektometer |
| 27,679,795 | kilograms per cubic dekameter |
| 27,679.7952 | kilograms per cubic meter |
| 783.809037 | kilograms per cubic foot |
| 0.453593 | kilograms per cubic inch |
| 27.679795 | kilograms per cubic decimeter |
| 0.0276798 | kilograms per cubic centimeter |
| 0.0000276798 | kilograms per cubic millimeter |

**POUND PER CUBIC INCH (cont'd): =**

61,023,360,960 . . . . . . . . pounds per cubic hektometer
61,023,361 . . . . . . . . pounds per cubic dekameter
61,023.360960 . . . . . . . . pounds per cubic meter
1,728 . . . . . . . . pounds per cubic foot
1 . . . . . . . . pounds per cubic inch
61.0233610 . . . . . . . . pounds per cubic decimeter
0.0610234 . . . . . . . . pounds per cubic centimeter
0.0000610234 . . . . . . . . pounds per cubic millimeter
276,797,952,000 . . . . . . . . hektograms per cubic hektometer
276,797,952 . . . . . . . . hektograms per cubic dekameter
276,798 . . . . . . . . hektograms per cubic meter
7,838.0903748 . . . . . . . . hektograms per cubic foot
4.535934 . . . . . . . . hektograms per cubic inch
276.797952 . . . . . . . . hektograms per cubic decimeter
0.276798 . . . . . . . . hektograms per cubic centimeter
0.000276798 . . . . . . . . hektograms per cubic millimeter
$276,797,952 \times 10^4$ . . . . . . . . dekagrams per cubic hektometer
2,767,979,520 . . . . . . . . dekagrams per cubic dekameter
2,767,980 . . . . . . . . dekagrams per cubic meter
78,380.903748 . . . . . . . . dekagrams per cubic foot
45.359342 . . . . . . . . dekagrams per cubic inch
2,767.97952 . . . . . . . . dekagrams per cubic decimeter
2.767980 . . . . . . . . dekagrams per cubic centimeter
0.00276798 . . . . . . . . dekagrams per cubic millimeter
976,373,775,360 . . . . . . . . ounces per cubic hektometer
975,373,775 . . . . . . . . ounces per cubic dekameter
976,374 . . . . . . . . ounces per cubic meter
27,648 . . . . . . . . ounces per cubic foot
16 . . . . . . . . ounces per cubic inch
976.373775 . . . . . . . . ounces per cubic decimeter
0.976374 . . . . . . . . ounces per cubic centimeter
0.000976374 . . . . . . . . ounces per cubic millimeter
$276,797,952 \times 10^5$ . . . . . . . . grams per cubic hektometer
27,679,795,200 . . . . . . . . grams per cubic dekameter
27,679,795 . . . . . . . . grams per cubic meter
783,809 . . . . . . . . grams per cubic foot
453.593425 . . . . . . . . grams per cubic inch
27,679.7952 . . . . . . . . grams per cubic decimeter
27.679795 . . . . . . . . grams per cubic centimeter
0.0276798 . . . . . . . . grams per cubic millimeter
$276,797,952 \times 10^6$ . . . . . . . . decigrams per cubic hektometer
$276,797,952 \times 10^3$ . . . . . . . . decigrams per cubic dekameter
276,797,952 . . . . . . . . decigrams per cubic meter
7,838,090 . . . . . . . . decigrams per cubic foot

POUND PER CUBIC INCH (cont'd): =

4,535.934249 . . . . . . . . . . decigrams per cubic inch
276,798 . . . . . . . . . . decigrams per cubic decimeter
276.797952 . . . . . . . . . . decigrams per cubic centimeter
0.276798 . . . . . . . . . . decigrams per cubic millimeter
$276,797,952 \times 10^7$ . . . . . . . . . . centigrams per cubic hektometer
$276,797,952 \times 10^4$ . . . . . . . . . . centigrams per cubic dekameter
2,767,979,520 . . . . . . . . . . centigrams per cubic meter
78,380,904 . . . . . . . . . . centigrams per cubic foot
45,359.342495 . . . . . . . . . . centigrams per cubic inch
2,767,980 . . . . . . . . . . centigrams per cubic decimeter
2,767,979520 . . . . . . . . . . centigrams per cubic centimeter
2.767980 . . . . . . . . . . centigrams per cubic millimeter
$276,797,952 \times 10^8$ . . . . . . . . . . milligrams per cubic hektometer
$276,797,952 \times 10^5$ . . . . . . . . . . milligrams per cubic dekameter
$276,797,952 \times 10^2$ . . . . . . . . . . milligrams per cubic meter
783,809,040 . . . . . . . . . . milligrams per cubic foot
453,593 . . . . . . . . . . milligrams per cubic inch
27,679.795 . . . . . . . . . . milligrams per cubic decimeter
27,679.7952 . . . . . . . . . . milligrams per cubic centimeter
27.679795 . . . . . . . . . . milligrams per cubic millimeter
$27,144,411 \times 10^9$ . . . . . . . . . . dynes per cubic hektometer
$27,144,411 \times 10^6$ . . . . . . . . . . dynes per cubic dekameter
$27,144,411 \times 10^3$ . . . . . . . . . . dynes per cubic meter
768,648,960 . . . . . . . . . . dynes per cubic foot
444,820 . . . . . . . . . . dynes per cubic inch
271,444,110 . . . . . . . . . . dynes per cubic decimeter
271,444 . . . . . . . . . . dynes per cubic centimeter
271.444411 . . . . . . . . . . dynes per cubic millimeter

POUND OF WATER EVAPORATED FROM AND AT 212° F.: =

24,271,651 . . . . . . . . . . foot poundals
291,259,807 . . . . . . . . . . inch poundals
754,402 . . . . . . . . . . foot pounds
9,052,822 . . . . . . . . . . inch pounds
0.244343 . . . . . . . . . . ton calories
244.343393 . . . . . . . . . . kilogram calories
538.685776 . . . . . . . . . . pound calories
2,443.433927 . . . . . . . . . . hektogram calories
24,434.339270 . . . . . . . . . . dekagram calories
8,618.949444 . . . . . . . . . . ounce calories
244,343 . . . . . . . . . . gram calories (mean)
2,443,434 . . . . . . . . . . decigram calories

| | |
|---|---|
| 24,434,339 | centigram calories |
| 244,343,393 | milligram calories |
| 0.0118359 | kilowatt days |
| 0.284061 | kilowatt hours |
| 17.0437064 | kilowatt minutes |
| 1,022.620366 | kilowatt seconds |
| 11.835871 | watt days |
| 284.0607707 | watt hours |
| 17,043.706381 | watt minutes |
| 1,022,620 | watt seconds |
| 104.32 | ton meters |
| 104,320 | kilogram meters |
| 229,942 | pound meters |
| 1,043,200 | hektogram meters |
| 10,432,000 | dekagram meters |
| 3,679,073 | ounce meters |
| 104,320,000 | gram meters |
| 1,043,200,000 | decigram meters |
| 10,432,000,000 | centigram meters |
| 104,320,000,000 | milligram meters |
| 1.0432 | ton hektometers |
| 1,043.2 | kilogram hektometers |
| 2,299.424641 | pounds hektometers |
| 10,432 | hektogram hektometers |
| 104,320 | dekagram hektometers |
| 36,790.753232 | ounce hektometers |
| 1,043,200 | gram hektometers |
| 10,432,000 | decigram hektometers |
| 104,320,000 | centigram hektometers |
| 1,043,200,000 | milligram hektometers |
| 10.432 | ton dekameters |
| 10,432 | kilogram dekameters |
| 22,994.24641 | pound dekameters |
| 104,320 | hektogram dekameters |
| 1,043,200 | dekagram dekameters |
| 367,907 | ounce dekameters |
| 10,432,000 | gram dekameters |
| 104,320,000 | decigram dekameters |
| 1,043,200,000 | centigram dekameters |
| 10,432,000,000 | milligram dekameters |
| 342.183304 | ton feet |
| 342,183 | kilogram feet |
| 754,271 | pound feet |
| 3,421,830 | hektogram feet |

## POUND OF WATER EVAPORATED FROM AND AT 212° F. (cont'd): =

| | |
|---|---|
| 34,218,300 | dekagram feet |
| 12,070,463 | ounce feet |
| 342,183,000 | gram feet |
| 3,421,830,000 | decigram feet |
| 34,218,300,000 | centigram feet |
| 342,183,000,000 | milligram feet |
| 4,106.298562 | ton inches |
| 4,106,299 | kilogram inches |
| 9,052,847 | pound inches |
| 41,062,986 | hektogram inches |
| 410,629,856 | dekagram inches |
| 144,845,540 | ounce inches |
| 4,106,298,562 | gram inches |
| 41,062,985,620 | decigram inches |
| 410,629,856,200 | centigram inches |
| 4,106,298,562 $\times 10_3$ | milligram inches |
| 1,043.2 | ton decimeters |
| 1,043,200 | kilogram decimeters |
| 2,299,425 | pound decimeters |
| 10,432,000 | hektogram decimeters |
| 104,320,000 | dekagram decimeters |
| 36,790,700 | ounce decimeters |
| 1,043,200,000 | gram decimeters |
| 10,432,000,000 | decigram decimeters |
| 104,320,000,000 | centigram decimeters |
| 10,432 $\times 10^8$ | milligram decimeters |
| 10,432 | ton centimeters |
| 10,432,000 | kilogram centimeters |
| 22,994250 | pound centimeters |
| 104,320,000 | hektogram centimeters |
| 1,043,200,000 | dekagram centimeters |
| 367,907,000 | ounce centimeters |
| 10,432,000,000 | gram centimeters |
| 104,320,000,000 | decigram centimeters |
| 10,432 $\times 10^8$ | centigram centimeters |
| 10,432 $\times 10^9$ | milligram centimeters |
| 104,320 | ton millimeters |
| 104,320,000 | kilogram millimeters |
| 229,942,500 | pound millimeters |
| 1,043,200,000 | hektogram millimeters |
| 10,432,000,000 | dekagram millimeters |
| 3,679,070,000 | ounce millimeters |
| 104,320,000,000 | gram millimeters |
| 10,432 $\times 10^8$ | decigram millimeters |

## POUND OF WATER EVAPORATED FROM AND AT 212° F. (cont'd): =

10,432 x $10^9$ . . . . . . . . . . centigram millimeters
10,432 x $10^{10}$ . . . . . . . . . . milligram millimeters
10.0938730 . . . . . . . . . . kiloliter-atmospheres
100.938730 . . . . . . . . . . hektoliter-atmospheres
1,009.387298 . . . . . . . . . . dekaliter-atmospheres
10,093.872982 . . . . . . . . . . liter-atmospheres
100,939 . . . . . . . . . . deciliter-atmospheres
1,009,387 . . . . . . . . . . centiliter-atmospheres
10,093,873 . . . . . . . . . . milliliter-atmospheres
0.0000000100939 . . . . . . . . . . cubic kilometer-atmospheres
0.0000100939 . . . . . . . . . . cubic hektometer-atmospheres
0.0100939 . . . . . . . . . . cubic dekameter-atmospheres
10.0938730 . . . . . . . . . . cubic meter-atmospheres
356.505500 . . . . . . . . . . cubic foot-atmospheres
615,979 . . . . . . . . . . cubic inch-atmospheres
10,093.872982 . . . . . . . . . . cubic decimeter-atmospheres
10,093,873 . . . . . . . . . . cubic centimeter-atmospheres
10,093,872,982 . . . . . . . . . . cubic millimeter-atmospheres
1,022,792 . . . . . . . . . . joules (absolute)
0.386293 . . . . . . . . . . Cheval-Vapeur hours
0.0158753 . . . . . . . . . . horsepower days
0.380993 . . . . . . . . . . horsepower hours
22.860503 . . . . . . . . . . horsepower minutes
1,371.626098 . . . . . . . . . . horsepower seconds
0.0664530 . . . . . . . . . . pounds of carbon oxidized with 100% efficiency
1 . . . . . . . . . . pounds of water evaporated from and at 212° F.
439,828 . . . . . . . . . . BTU (mean) per pound
969.718377 . . . . . . . . . . BTU (mean

## POUND OF WATER AT 60° F.: =

0.0000454248 . . . . . . . . . . kiloliters
0.0000454248 . . . . . . . . . . cubic meters
0.00059412 . . . . . . . . . . cubic yards
0.0028572 . . . . . . . . . . barrels
0.00454248 . . . . . . . . . . hektoliters
0.0128904 . . . . . . . . . . bushels (U.S.—dry)
0.0124968 . . . . . . . . . . bushels (Imperial—dry)
0.0160417 . . . . . . . . . . cubic feet
0.0454248 . . . . . . . . . . dekaliters
0.12 . . . . . . . . . . gallons (U.S.—liquid)
0.103138 . . . . . . . . . . gallons (U.S.—dry)
0.0999216 . . . . . . . . . . gallons (Imperial)

POUND OF WATER AT 60° F. (cont'd): =

0.48 . . . . . . . . . . quarts (liquid)
0.412496 . . . . . . . . . . quarts (dry)
0.454248 . . . . . . . . . . liters
0.454248 . . . . . . . . . . cubic decimeters
0.96 . . . . . . . . . . pints
3.84 . . . . . . . . . . gills
4.54248 . . . . . . . . . . deciliters
27.72 . . . . . . . . . . cubic inches
45.4248 . . . . . . . . . . centiliters
454.248 . . . . . . . . . . milliliters
454.248 . . . . . . . . . . cubic centimeters
454,248 . . . . . . . . . . cubic millimeters
7,372.80 . . . . . . . . . . minims
15.36 . . . . . . . . . . ounces (fluid)
122.88 . . . . . . . . . . drams (fluid)
0.000446628 . . . . . . . . . . tons (long)
0.000453792 . . . . . . . . . . tons (metric)
0.000500220 . . . . . . . . . . tons (short)
0.453792 . . . . . . . . . . kilograms
1 . . . . . . . . . . pounds (Avoir.)
1.215812 . . . . . . . . . . pounds (Troy)
4.53792 . . . . . . . . . . hektograms
45.3792 . . . . . . . . . . dekagrams
14.589749 . . . . . . . . . . ounces (Troy)
16 . . . . . . . . . . ounces (Avoir.)
116.717990 . . . . . . . . . . drams (Troy)
256 . . . . . . . . . . drams (Avoir.)
35.0153971 . . . . . . . . . . scruples
453.792 . . . . . . . . . . grams
4,537.92 . . . . . . . . . . decigrams
7,000 . . . . . . . . . . grains
45,379.2 . . . . . . . . . . centigrams
453,792 . . . . . . . . . . milligrams
27.6798 . . . . . . . . . . inches of water

POUNDS PER MILLION GALLONS: =

0.000000132090 . . . . . . . . . . tons (net) per cubic meter
0.00011983 . . . . . . . . . . kilograms per cubic meter
0.000264180 . . . . . . . . . . pounds (Avoir.) per cubic meter
0.00422687 . . . . . . . . . . ounces (Avoir.) per cubic meter
0.11983 . . . . . . . . . . grams per cubic meter
1.849264 . . . . . . . . . . grains per cubic meter

POUND PER MILLION GALLONS (cont'd): =

0.0000000210002 . . . . . . . . . . . . . . . . . . . . . . . . . . tons (net) per barrel
0.0000190510 . . . . . . . . . . . . . . . . . . . . . . . . . . kilograms per barrel
0.0000420003 . . . . . . . . . . . . . . . . . . . . . . pounds (Avoir.) per barrel
0.000672005 . . . . . . . . . . . . . . . . . . . . . . . . ounces (Avoir.) per barrel
0.0190510 . . . . . . . . . . . . . . . . . . . . . . . . . . . . . . grams per barrel
0.294004 . . . . . . . . . . . . . . . . . . . . . . . . . . . . . . . grains per barrel
0.00000000370029 . . . . . . . . . . . . . . . . . . . . tons (net) per cubic foot
0.00000339313 . . . . . . . . . . . . . . . . . . . . . . . kilograms per cubic foot
0.00000748056 . . . . . . . . . . . . . . . . . . . pounds (Avoir.) per cubic foot
0.000119689 . . . . . . . . . . . . . . . . . . . . . ounces (Avoir.) per cubic foot
0.00339313 . . . . . . . . . . . . . . . . . . . . . . . . . . . grams per cubic foot
0.0523642 . . . . . . . . . . . . . . . . . . . . . . . . . . . . grains per cubic foot
0.0000000005 . . . . . . . . . . . . . . . . . tons (net) per gallon (U.S.—liquid)
0.000000453585 . . . . . . . . . . . . . . . kilograms per gallon (U.S.—liquid)
0.000001 . . . . . . . . . . . . . . . . . . . pounds (Avoir.) per gallon (U.S.—liquid)
0.000016 . . . . . . . . . . . . . . . . . . . ounces (Avoir.) per gallon (U.S.—liquid)
0.000453585 . . . . . . . . . . . . . . . . . . . . grams per gallon (U.S.—liquid)
0.007 . . . . . . . . . . . . . . . . . . . . . . . . . . . grains per gallon (U.S.—liquid)
0.000000000600477 . . . . . . . . . . tons (net) per gallon (Imperial—liquid)
0.000000544729 . . . . . . . . . . . . kilograms per gallon (Imperial—liquid)
0.00000120095 . . . . . . . . . . . . pounds (Avoir.) per gallon (Imp.—liquid)
0.0000192153 . . . . . . . . . . . . . ounces (Avoir.) per gallon (Imp.—liquid)
0.000544729 . . . . . . . . . . . . . . . . . grams per gallon (Imperial—liquid)
0.00840670 . . . . . . . . . . . . . . . . . grains per gallon (Imperial—liquid)
0.000000000132090 . . . . . . . . . . . . . . . . . . . . . . . tons (net) per liter
0.00000011983 . . . . . . . . . . . . . . . . . . . . . . . . . . kilograms per liter
0.000000264180 . . . . . . . . . . . . . . . . . . . . . pounds (Avoir.) per liter
0.00000422687 . . . . . . . . . . . . . . . . . . . . . . ounces (Avoir.) per liter
0.00011983 . . . . . . . . . . . . . . . . . . . . . . . . . . . . . . . grams per liter
0.00184926 . . . . . . . . . . . . . . . . . . . . . . . . . . . . . . grains per liter
0.00000000000216453 . . . . . . . . . . . . . . . . . tons (net) per cubic inch
0.00000000196362 . . . . . . . . . . . . . . . . . . . kilograms per cubic inch
0.00000000432903 . . . . . . . . . . . . . . . . . pounds (Avoir.) per cubic inch
0.0000000692645 . . . . . . . . . . . . . . . . . . ounces (Avoir.) per cubic inch
0.00000196362 . . . . . . . . . . . . . . . . . . . . . . . . grams per cubic inch
0.0000303033 . . . . . . . . . . . . . . . . . . . . . . . . . grains per cubic inch

POUND WEIGHT SECOND PER SQUARE FOOT: =

478.8 . . . . . . . . . . . . . . . . . . . . . . . . . . . . . . . . . . . . . . . . poises

## POUND WEIGHT PER SECOND PER SQUARE INCH: =

68,950 . . . . . . . . . . . . . . . . . . . . . . . . . . . . . . . . . . . . . . . poises

## QUADRANT (ANGLE): =

324,000 . . . . . . . . . . . . . . . . . . . . . . . . . . . . . . . . . . . . . seconds
5,400 . . . . . . . . . . . . . . . . . . . . . . . . . . . . . . . . . . . . . . minutes
90 . . . . . . . . . . . . . . . . . . . . . . . . . . . . . . . . . . . . . . . . degrees
1.57080 . . . . . . . . . . . . . . . . . . . . . . . . . . . . . . . . . . . . . radians
0.25 . . . . . . . . . . . . . . . . . . . . . . . . . . circumference or revolution
0.7854 . . . . . . . . . . . . . . . . . . . . . . . . . . . . . . . . . . . . . . Pi ($\pi$)

## QUART (U.S.–Dry): =

0.000110089 . . . . . . . . . . . . . . . . . . . . . . . . . . . . . . kiloliters
0.000110089 . . . . . . . . . . . . . . . . . . . . . . . . . . . . cubic meters
0.00143986 . . . . . . . . . . . . . . . . . . . . . . . . . . . . . cubic yards
0.00692448 . . . . . . . . . . . . . . . . . . . . . . . . . . . . . . . . barrels
0.0110089 . . . . . . . . . . . . . . . . . . . . . . . . . . . . . . hektoliters
0.0312402 . . . . . . . . . . . . . . . . . . . . . . . . . bushels (U.S.–dry)
0.0302863 . . . . . . . . . . . . . . . . . . . . . . . bushels (Imperial–dry)
0.0388775 . . . . . . . . . . . . . . . . . . . . . . . . . . . . . . . cubic feet
0.110089 . . . . . . . . . . . . . . . . . . . . . . . . . . . . . . . dekaliters
0.290823 . . . . . . . . . . . . . . . . . . . . . . . . . gallons (U.S.–liquid)
0.249956 . . . . . . . . . . . . . . . . . . . . . . . . . . .gallons (U.S.–dry)
0.242162 . . . . . . . . . . . . . . . . . . . . . . . . . . . gallons (Imperial)
1.163290 . . . . . . . . . . . . . . . . . . . . . . . . . . . . . quarts (liquid)
1 . . . . . . . . . . . . . . . . . . . . . . . . . . . . . . . . . . . . .quarts (dry)
1.100889 . . . . . . . . . . . . . . . . . . . . . . . . . . . . . . . . . . . .liters
1.100889 . . . . . . . . . . . . . . . . . . . . . . . . . . . cubic decimeters
2 . . . . . . . . . . . . . . . . . . . . . . . . . . . . . . . . . . pints (U.S.–dry)
2.326580 . . . . . . . . . . . . . . . . . . . . . . . . . . pints (U.S.–liquid)
7.749187 . . . . . . . . . . . . . . . . . . . . . . . . . . . . . gills (Imperial)
9.306320 . . . . . . . . . . . . . . . . . . . . . . . . . . . . . . . . gills (U.S.)
11.00888839 . . . . . . . . . . . . . . . . . . . . . . . . . . . . . . . deciliters
67.18 . . . . . . . . . . . . . . . . . . . . . . . . . . . . . . . . . cubic inches
110.0888839 . . . . . . . . . . . . . . . . . . . . . . . . . . . . . . . centiliters
1,100.888839 . . . . . . . . . . . . . . . . . . . . . . . . . . . . . . milliliters
1,100.888839 . . . . . . . . . . . . . . . . . . . . . . . . . cubic centimeters
1,100,889 . . . . . . . . . . . . . . . . . . . . . . . . . . . .cubic millimeters
17,868.135060 . . . . . . . . . . . . . . . . . . . . . . . . . . . . . . . minims
37.225281 . . . . . . . . . . . . . . . . . . . . . . . . . . . . . ounces (fluid)
297.802251 . . . . . . . . . . . . . . . . . . . . . . . . . . . . . drams (fluid)

QUART (U.S.–DRY) (cont'd): =

| | |
|---|---|
| 0.00108241 | tons (long) |
| 0.00109977 | tons (metric) |
| 0.00121230 | tons (short) |
| 1.0997744 | kilograms |
| 2.424587 | pounds (Avoir.) |
| 2.946547 | pounds (Troy) |
| 10.997744 | hektograms |
| 109.977441 | dekagrams |
| 35.358561 | ounces (Troy) |
| 38.793396 | ounces (Avoir.) |
| 282.868492 | drams (Troy) |
| 620.694342 | drams (Avoir.) |
| 707.171230 | pennyweights |
| 848.605475 | scruples |
| 1,099.774407 | grams |
| 10,997.744067 | decigrams |
| 16,972.110905 | grains |
| 109,977 | centigrams |
| 1,099,774 | milligrams |

QUART (U.S.–LIQUID): =

| | |
|---|---|
| 0.0000946358 | kiloliters |
| 0.0000946358 | cubic meters |
| 0.00123775 | cubic yards |
| 0.0059525 | barrels |
| 0.00946358 | hektoliters |
| 0.026855 | bushels (U.S.–dry) |
| 0.026035 | bushels (Imperial–dry) |
| 0.0334203 | cubic feet |
| 0.0946358 | dekaliters |
| 0.25 | gallons (U.S.–liquid) |
| 0.21487 | gallons (U.S.–dry) |
| 0.20817 | gallons (Imperial) |
| 1 | quarts (liquid) |
| 0.859368 | quarts (dry) |
| 0.946358 | liters |
| 0.946358 | cubic decimeters |
| 1.718733 | pints (U.S.–dry) |
| 2 | pints (U.S.–liquid) |
| 6.66144 | gills (Imperial) |
| 8 | gills (U.S.) |
| 9.46358 | deciliters |

**QUART (U.S.–LIQUID) (cont'd): =**

| | |
|---|---|
| 57.75 | cubic inches |
| 94.6358 | centiliters |
| 946.358 | milliliters |
| 946.358 | cubic centimeters |
| 946,358 | cubic millimeters |
| 15,360 | minims |
| 32 | ounces (fluid) |
| 256 | drams (fluid) |
| 0.000930475 | tons (long) water @ 62° F. |
| 0.0009454 | tons (metric) water @ 62° F. |
| 0.00104213 | tons (short) water @ 62° F. |
| 0.9454 | kilograms water @ 62° F. |
| 2.08425 | pounds (Avoir.) water @ 62° F. |
| 2.532943 | pounds (Troy) water @ 62° F. |
| 9.455 | hektograms water @ 62° F. |
| 94.55 | dekagrams water @ 62° F. |
| 30.39531 | ounces (Troy) water @ 62° F. |
| 33.348 | ounces (Avoir.) water @ 62° F. |
| 243.16248 | drams (Troy) water @ 62° F. |
| 533.568 | drams (Avoir.) water @ 62° F. |
| 607.9062 | pennyweights water @ 62° F. |
| 729.48744 | scruples water @ 62° F. |
| 945.5 | grams water @ 62° F. |
| 9,455 | decigrams water @ 62° F. |
| 14,589.75 | grains water @ 62° F. |
| 94,550 | centigrams water @ 62° F. |
| 945,500 | milligrams water @ 62° F. |

**QUIRE: =**

25 sheets

**RADIAN: =**

| | |
|---|---|
| 206,265 | seconds or inches |
| 3,437.75 | minutes |
| 57.29578 | degrees |
| 0.637 | quadrants |
| 0.159155 | circumference or revolutions |
| 0.5 | Pi ($\pi$) |
| 57° 17′ 44.8″ | (In degrees, minutes, and seconds) |

## RADIANS PER SECOND: =

| | |
|---|---|
| 57.29578 | degrees per second |
| 3,437.7468 | degrees per minute |
| 206,265 | degrees per hour |
| 4,950.355 | degrees per day |
| 0.637 | quadrants per second |
| 38.22 | quadrants per minute |
| 2,293.2 | quadrants per hour |
| 55,036.8 | quadrants per day |
| 0.159155 | revolutions per second |
| 9.5493 | revolutions per minute |
| 572.958 | revolutions per hour |
| 13,750.992 | revolutions per day |

## RADIAN PER SECOND PER SECOND: =

| | |
|---|---|
| 57.29578 | degrees per second per second |
| 3,437.7468 | degrees per minute per second |
| 206,265 | degrees per minute per minute |
| 0.637 | quadrants per second per second |
| 38.22 | quadrants per minute per second |
| 2,293.2 | quadrants per minute per minute |
| 0.159155 | revolutions per second per second |
| 9.5493 | revolutions per minute per second |
| 572.958 | revolutions per minute per minute |

## REAM: =

| | |
|---|---|
| 500 | sheets |
| 20 | quires |

## REVOLUTION: =

| | |
|---|---|
| 1,296,000 | seconds or inches |
| 21,600 | minutes |
| 360 | degrees |
| 6.2832 | radians |
| 4 | quadrants |
| 2 | Pi ($\pi$) |
| 1 | circumference |

REVOLUTIONS PER SECOND PER SECOND: =

| | |
|---|---|
| 360 | degrees per second per second |
| 21,600 | degrees per minute per second |
| 1,296,000 | degrees per minute per minute |
| 6.2832 | radians per second per second |
| 376.9920 | radians per minute per second |
| 22,619.52 | radians per minute per minute |
| 4 | quadrants per second per second |
| 240 | quadrants per minute per second |
| 14,400 | quadrants per minute per minute |
| 1 | revolutions per second per second |
| 60 | revolutions per minute per second |
| 3,600 | revolutions per minute per minute |

REVOLUTIONS PER MINUTE PER MINUTE: =

| | |
|---|---|
| 0.1 | degrees per second per second |
| 6 | degrees per minute per second |
| 360 | degrees per minute per minute |
| 0.0017453 | radians per second per second |
| 0.104718 | radians per minute per second |
| 6.2832 | radians per minute per minute |
| 0.00111111 | quadrants per second per second |
| 0.06666667 | quadrants per minute per second |
| 4 | quadrants per minute per minute |
| 0.000277778 | revolutions per second per second |
| 0.0166667 | revolutions per minute per second |
| 1 | revolutions per minute per minute |

REVOLUTIONS PER SECOND: =

| | |
|---|---|
| 111,974,400,000 | seconds or inches per day |
| 4,665,600,000 | seconds or inches per hour |
| 77,760,000 | seconds or inches per minute |
| 1,296,000 | seconds or inches per second |
| 1,866,240,000 | minutes per day |
| 77,760,000 | minutes per hour |
| 1,296,000 | minutes per minute |
| 21,600 | minutes per second |
| 31,104,000 | degrees per day |
| 1,296,000 | degrees per hour |
| 21,600 | degrees per minute |
| 360 | degrees per second |

## REVOLUTIONS PER SECOND (cont'd): =

| | |
|---|---|
| 542,868 | radians per day |
| 22,619.52 | radians per hour |
| 376.9920 | radians per minute |
| 6.2832 | radians per second |
| 345,600 | quadrants per day |
| 14,400 | quadrants per hour |
| 240 | quadrants per minute |
| 4 | quadrants per second |
| 172,800 | Pi ($\pi$) per day |
| 7,200 | Pi ($\pi$) per hour |
| 120 | Pi ($\pi$) per minute |
| 2 | Pi ($\pi$) per second |
| 86,400 | revolutions or circumferences per day |
| 3,600 | revolutions or circumferences per hour |
| 60 | revolutions or circumferences per minute |
| 1 | revolutions or circumferences per second |

## REVOLUTION PER MINUTE: =

| | |
|---|---|
| 1,866,240,000 | seconds or inches per day |
| 77,760,000 | seconds or inches per hour |
| 1,296,000 | seconds or inches per minute |
| 21,600 | seconds or inches per second |
| 31,104,00 | minutes per day |
| 1,296,000 | minutes per hour |
| 21,600 | minutes per minute |
| 360 | minutes per second |
| 518,400 | degrees per day |
| 21,600 | degrees per hour |
| 360 | degrees per minute |
| 6 | degrees per second |
| 9,047.808 | radians per day |
| 376.992 | radians per hour |
| 6.2832 | radians per minute |
| 0.10472 | radians per second |
| 5,760 | quadrants per day |
| 240 | quadrants per hour |
| 4 | quadrants per minute |
| 0.0666667 | quadrants per second |
| 2,880 | Pi ($\pi$) per day |
| 120 | Pi ($\pi$) per hour |
| 2 | Pi ($\pi$) per minute |
| 0.0333333 | Pi ($\pi$) per second |

REVOLUTION PER MINUTE (cont'd): =

| | |
|---|---|
| 1,440 | revolutions or circumferences per day |
| 60 | revolutions or circumferences per hour |
| 1 | revolutions or circumferences per minute |
| 0.016667 | revolutions or circumferences per second |

REVOLUTION PER HOUR: =

| | |
|---|---|
| 31,104,000 | seconds or inches per day |
| 1,296,000 | seconds or inches per hour |
| 21,600 | seconds or inches per minute |
| 360 | seconds or inches per second |
| 518,400 | minutes per day |
| 21,600 | minutes per hour |
| 360 | minutes per minute |
| 6 | minutes per second |
| 8,640 | degrees per day |
| 360 | degrees per hour |
| 6 | degrees per minute |
| 0.1 | degrees per second |
| 150.796512 | radians per day |
| 6.2832 | radians per hour |
| 0.104720 | radians per minute |
| 0.00174533 | radians per second |
| 96 | quadrants per day |
| 4 | quadrants per hour |
| 0.0666667 | quadrants per minute |
| 0.00111111 | quadrants per second |
| 48 | Pi ($\pi$) per day |
| 2 | Pi ($\pi$) per hour |
| 0.0333333 | Pi ($\pi$) per minute |
| 0.000555556 | Pi ($\pi$) per second |
| 24 | revolutions or circumferences per day |
| 1 | revolutions or circumferences per hour |
| 0.0166667 | revolutions or circumferences per minute |
| 0.000277778 | revolutions or circumferences per second |

REVOLUTIONS PER DAY: =

| | |
|---|---|
| 1,296,000 | seconds or inches per day |
| 54,000 | seconds or inches per hour |
| 900 | seconds or inches per minute |
| 15 | seconds or inches per second |

## REVOLUTIONS PER DAY (cont'd): =

| | |
|---|---|
| 21,600 | minutes per day |
| 900 | minutes per hour |
| 15 | minutes per minute |
| 0.25 | minutes per second |
| 360 | degrees per day |
| 15 | degrees per hour |
| 0.25 | degrees per minute |
| 0.00416667 | degrees per second |
| 6.2832 | radians per day |
| 0.2618 | radians per hour |
| 0.00436333 | radians per minute |
| 0.0000727222 | radians per second |
| 4 | quadrants per day |
| 0.166667 | quadrants per hour |
| 0.00277778 | quadrants per minute |
| 0.0000462963 | quadrants per second |
| 2 | Pi ($\pi$) per day |
| 0.0833333 | Pi ($\pi$) per hour |
| 0.00138889 | Pi ($\pi$) per minute |
| 0.0000231481 | Pi ($\pi$) per second |
| 1 | revolutions or circumferences per day |
| 0.0416667 | revolutions or circumferences per hour |
| 0.000694446 | revolutions or circumferences per minute |
| 0.0000115741 | revolutions or circumferences per second |

## ROD: =

| | |
|---|---|
| 0.00271363 | miles (nautical) |
| 0.003125 | miles (statute) |
| 0.00502922 | kilometers |
| 0.025 | furlongs |
| 0.0502922 | hektometers |
| 0.25 | chains |
| 0.502922 | dekameters |
| 1 | rods |
| 5.0292188 | meters |
| 5.5 | yards |
| 5.94 | varas (Texas) |
| 16.5 | feet |
| 22 | spans |
| 25 | links |
| 49.5 | hands |
| 50.292188 | decimeters |

**ROD (cont'd): =**

| | |
|---|---|
| 502.921875 | centimeters |
| 198 | inches |
| 5,029.21875 | millimeters |
| 198,000 | mils |
| 5,029,219 | microns |
| 50,029,218,750 | milli-microns |
| 5,029,218,750 | micro-millimeters |
| 7,811,165 | wave lengths of red line of cadmium |
| 50,292,187,500 | Angstrom units |

**SACK CEMENT: =**

| | |
|---|---|
| 0.19592 | barrels |
| 94 | pounds (Avoir.) |
| 8.22857 | gallons (U.S.—liquid) |
| 1.1 | cubic feet (set) |
| 1,900.8 | cubic inches |
| 3.15 | specific gravity |
| 0.484 | cubic feet (absolute volume) |

**SEAWATER GRAVITY: =**

1.02 to 1.03

**SECOND, FOOT (WATER): =**

| | |
|---|---|
| 2,446.594024 | kiloliters per day |
| 101.941418 | kiloliters per hour |
| 1.699024 | kiloliters per minute |
| 0.0283171 | kiloliters per second |
| 2,446.594024 | cubic meters per day |
| 101.941418 | cubic meters per hour |
| 1.699024 | cubic meters |
| 0.0283171 | cubic meters per second |
| 3,200.144983 | cubic yards per day |
| 133.339374 | cubic yards per hour |
| 2.222323 | cubic yards per minute |
| 0.0370378 | cubic yards per second |
| 15,388.959079 | barrels per day |
| 641.206628 | barrels per hour |
| 10.686777 | barrels per minute |

SECOND, FOOT (WATER) (cont'd): =

0.178113 . . . . . . . . . . . . . . . . . . . . . . . . . . . . . . . barrels per second
24,465.940237 . . . . . . . . . . . . . . . . . . . . . . . .hektoliters per day
1,019.414177 . . . . . . . . . . . . . . . . . . . . . . . . hektoliters per hour
16.990236 . . . . . . . . . . . . . . . . . . . . . . . . .hektoliters per minute
0.283171 . . . . . . . . . . . . . . . . . . . . . . . . . .hektoliters per second
69,429.445206 . . . . . . . . . . . . . . . . . . .bushels (U.S.—dry) per day
2,892.893550 . . . . . . . . . . . . . . . . . . . bushels (U.S.—dry) per hour
48.214893 . . . . . . . . . . . . . . . . . . . . . .bushels (U.S.—dry) per minute
0.803582 . . . . . . . . . . . . . . . . . . . . . . . bushels (U.S.—dry) per second
67,273.995871 . . . . . . . . . . . . . . . . . . bushels (Imperial—dry) per day
2,803.0831613 . . . . . . . . . . . . . . . . . .bushels (Imperial—dry) per hour
46.718053 . . . . . . . . . . . . . . . . . . . bushels (Imperial—dry) per minute
0.778634 . . . . . . . . . . . . . . . . . . . . bushels (Imperial—dry) per second
86,400 . . . . . . . . . . . . . . . . . . . . . . . . . . . . . . cubic feet per day
3,600 . . . . . . . . . . . . . . . . . . . . . . . . . . . . . . cubic feet per hour
60 . . . . . . . . . . . . . . . . . . . . . . . . . . . . . . . cubic feet per minute
1 . . . . . . . . . . . . . . . . . . . . . . . . . . . . . . . . cubic feet per second
244,659 . . . . . . . . . . . . . . . . . . . . . . . . . . . . . . dekaliters per day
10,194.141766 . . . . . . . . . . . . . . . . . . . . . . . . .dekaliters per hour
169,902363 . . . . . . . . . . . . . . . . . . . . . . . . . . dekaliters per minute
2,831706 . . . . . . . . . . . . . . . . . . . . . . . . . . . dekaliters per second
277,718 . . . . . . . . . . . . . . . . . . . . . . . . .pecks (U.S.—dry) per day
11,571.574201 . . . . . . . . . . . . . . . . . . . . pecks (U.S.—dry) per hour
192.859570 . . . . . . . . . . . . . . . . . . . . .pecks (U.S.—dry) per minute
3,214326 . . . . . . . . . . . . . . . . . . . . . . .pecks (U.S.—dry) per second
646,336 . . . . . . . . . . . . . . . . . . . . . . .gallons (U.S.—liquid) per day
26,930.679834 . . . . . . . . . . . . . . . . . . gallons (U.S.—liquid) per hour
448.844664 . . . . . . . . . . . . . . . . . . . .gallons (U.S.—liquid) per minute
7,480744 . . . . . . . . . . . . . . . . . . . . . .gallons (U.S.—liquid) per second
555,426 . . . . . . . . . . . . . . . . . . . . . . . . . gallons (U.S.—dry) per day
23,142.740636 . . . . . . . . . . . . . . . . . . . . gallons (U.S.—dry) per hour
385.712344 . . . . . . . . . . . . . . . . . . . . . gallons (U.S.—dry) per minute
6.428539 . . . . . . . . . . . . . . . . . . . . . . . gallons (U.S.—dry) per second
538,202 . . . . . . . . . . . . . . . . . . . . . . . . . .gallons (Imperial) per day
22,425.0730560 . . . . . . . . . . . . . . . . . . . . gallons (Imperial) per hour
373.751218 . . . . . . . . . . . . . . . . . . . . . .gallons (Imperial) per minute
6.229187 . . . . . . . . . . . . . . . . . . . . . . . .gallons (Imperial) per second
2,585,340 . . . . . . . . . . . . . . . . . . . . . . . . . . quarts (liquid) per day
107,723 . . . . . . . . . . . . . . . . . . . . . . . . . . . quarts (liquid) per hour
1,795.375258 . . . . . . . . . . . . . . . . . . . . . . quarts (liquid) per minute
29.922921 . . . . . . . . . . . . . . . . . . . . . . . .quarts (liquid) per second
2,221,757 . . . . . . . . . . . . . . . . . . . . . . . . . . . .quarts (dry) per day
92,573.205256 . . . . . . . . . . . . . . . . . . . . . . . quarts (dry) per hour
1,542.886754 . . . . . . . . . . . . . . . . . . . . . . . .quarts (dry) per minute

SECOND, FOOT (WATER) (cont'd): =

25.714779 . . . . . . . . . . . . . . . . . . . . . . . . . . . . quarts (dry) per second
2,446,594 . . . . . . . . . . . . . . . . . . . . . . . . . . . . . liters per day
101,941 . . . . . . . . . . . . . . . . . . . . . . . . . . . . . liters per hour
1,699.0236276 . . . . . . . . . . . . . . . . . . . . . . . . . liters per minute
28.317060 . . . . . . . . . . . . . . . . . . . . . . . . . . liters per second
2,446,594 . . . . . . . . . . . . . . . . . . . . . . . . cubic decimeters per day
101,941 . . . . . . . . . . . . . . . . . . . . . . . . cubic decimeters per hour
1,699.0236276 . . . . . . . . . . . . . . . . . . cubic decimeters per minute
28.317060 . . . . . . . . . . . . . . . . . . . . . cubic decimeters per second
4,443,513 . . . . . . . . . . . . . . . . . . . . . . . . pints (U.S.–dry) per day
185,146 . . . . . . . . . . . . . . . . . . . . . . . . pints (U.S.–dry) per hour
3,085.766712 . . . . . . . . . . . . . . . . . . . . pints (U.S.–dry) per minute
51.429445 . . . . . . . . . . . . . . . . . . . . . . pints (U.S.–dry) per second
5,170,639 . . . . . . . . . . . . . . . . . . . . . . pints (U.S.–liquid) per day
215,443 . . . . . . . . . . . . . . . . . . . . . . pints (U.S.–liquid) per hour
3,590.716535 . . . . . . . . . . . . . . . . . pints (U.S.–liquid) per minute
59.845276 . . . . . . . . . . . . . . . . . . . pints (U.S.–liquid) per second
17,222,068 . . . . . . . . . . . . . . . . . . . . . . gills (Imperial) per day
717,586 . . . . . . . . . . . . . . . . . . . . . . gills (Imperial) per hour
11,959.767119 . . . . . . . . . . . . . . . . . . gills (Imperial) per minute
199.329452 . . . . . . . . . . . . . . . . . . . . gills (Imperial) per second
20,683,007 . . . . . . . . . . . . . . . . . . . . . . . . . gills (U.S.) per day
861,792 . . . . . . . . . . . . . . . . . . . . . . . . . gills (U.S.) per hour
14,363.205943 . . . . . . . . . . . . . . . . . . . . gills (U.S.) per minute
239.386766 . . . . . . . . . . . . . . . . . . . . . . gills (U.S.) per second
24,465,940 . . . . . . . . . . . . . . . . . . . . . . . . . deciliters per day
1,019,414 . . . . . . . . . . . . . . . . . . . . . . . . . deciliters per hour
16,990.236276 . . . . . . . . . . . . . . . . . . . . . . deciliters per minute
283.170605 . . . . . . . . . . . . . . . . . . . . . . . deciliters per second
149,303,400 . . . . . . . . . . . . . . . . . . . . . . cubic inches per day
6,220,975 . . . . . . . . . . . . . . . . . . . . . . . cubic inches per hour
103,683 . . . . . . . . . . . . . . . . . . . . . . . cubic inches per minute
1,728.048146 . . . . . . . . . . . . . . . . . . . . cubic inches per second
244,659,402 . . . . . . . . . . . . . . . . . . . . . . . . centiliters per day
10,194,142 . . . . . . . . . . . . . . . . . . . . . . . . centiliters per hour
169,902 . . . . . . . . . . . . . . . . . . . . . . . . centiliters per minute
2,831.706046 . . . . . . . . . . . . . . . . . . . . . . centiliters per second
2,446,594,024 . . . . . . . . . . . . . . . . . . . . . . . milliliters per day
101,941,418 . . . . . . . . . . . . . . . . . . . . . . . milliliters per hour
1,699,023 . . . . . . . . . . . . . . . . . . . . . . . . milliliters per minute
28,317.06046 . . . . . . . . . . . . . . . . . . . . . . milliliters per second
2,446,594,024 . . . . . . . . . . . . . . . . . . . cubic centimeters per day
101,941,418 . . . . . . . . . . . . . . . . . . . . cubic centimeters per hour
1,699,023 . . . . . . . . . . . . . . . . . . . . . . cubic centimeters per minute

SECOND, FOOT (WATER) (cont'd): =

| | |
|---|---|
| 28,317.06046 | cubic centimeters per second |
| 2,446,594,024 x $10^3$ | cubic millimeters per day |
| 101,941,417,656 | cubic millimeters per hour |
| 1,699,023,627 | cubic millimeters per minute |
| 28,317,060 | cubic millimeters per second |
| 82,730,841 | ounces (U.S.) fluid per day |
| 3,447,121 | ounces (U.S.) fluid per hour |
| 57,451.974260 | ounces (U.S.) fluid per minute |
| 957.532904 | ounces (U.S.) fluid per second |
| 86,110,312 | ounces (Imperial—fluid) per day |
| 3,587,941 | ounces (Imperial—fluid) per hour |
| 59,798.835597 | ounces (Imperial—fluid) per minute |
| 996.647260 | ounces (Imperial—fluid) per second |
| 661,847,490 | drams (fluid) per day |
| 27,576,966 | drams (fluid) per hour |
| 459,616 | drams (fluid) per minute |
| 7,660.271730 | drams (fluid) per second |
| 39,710,848,658 | minims per day |
| 1,654,618,682 | minims per hour |
| 27,576,966 | minims per minute |
| 459,616 | minims per second |
| 1.98347 | acre feet |
| 1 | acre, inch per hour |

SECOND, (ANGLE): =

| | |
|---|---|
| 1 | seconds |
| 0.0166667 | minutes |
| 0.000277778 | degrees |
| 0.00000484814 | radians |
| 0.00000308651 | quadrants |
| 0.00000154321 | Pi ($\pi$) |
| 0.000000771607 | circumferences or revolutions |

SQUARE CENTIMETER: =

| | |
|---|---|
| 0.00000000003831 | square miles or sections |
| 0.0000000001 | square kilometers |
| 0.00000000247104 | square furlongs |
| 0.0000000247104 | acres |
| 0.00000001 | square hektometers or hectares |
| 0.000000247104 | square chains |

SQUARE CENTIMETER (cont'd): =

| | |
|---|---|
| 0.000001 | square dekameters or acres |
| 0.00000395367 | square rods |
| 0.0001 | square meters or centares |
| 0.00011960 | square yards |
| 0.000139498 | square varas (Texas) |
| 0.00107639 | square feet |
| 0.00247104 | square links |
| 0.01 | square decimeters |
| 1 | square centimeters |
| 0.1550 | square inches |
| 100 | square millimeters |
| 155,000 | square mils |
| 197,350 | circular mils |
| 127.32 | circular millimeters |

SQUARE CHAIN: =

| | |
|---|---|
| 0.00015625 | square miles or sections |
| 0.000404687 | square kilometers |
| 0.01 | furlongs |
| 0.1 | acres |
| 0.0404687 | square hektometers or hectares |
| 1 | square chains |
| 4.046873 | square dekameters or ares |
| 16 | square rods |
| 404.6873 | square meters or centares |
| 484 | square yards |
| 564.530690 | square varas (Texas) |
| 4,356 | square feet |
| 10,000 | square links |
| 40,468.73 | square decimeters |
| 4,046,873 | square centimeters |
| 627,265 | square inches |
| 404,687,300 | square millimeters |
| 627,265,000,000 | square mils |
| 798,650,386,550 | circular mils |
| 515,247,870 | circular millimeters |

SQUARE DECIMETER: =

| | |
|---|---|
| 0.000000003831 | square miles or sections |
| 0.00000001 | square kilometers |

**SQUARE DECIMETER (cont'd): =**

0.000000247104 . . . . . . . . . . . . . . . . . . . . . . . . . . .square furlongs
0.00000247104 . . . . . . . . . . . . . . . . . . . . . . . . . . . . . . . . .acres
0.000001 . . . . . . . . . . . . . . . . . . . . square hektometers or hectares
0.0000247104 . . . . . . . . . . . . . . . . . . . . . . . . . . . . . .square chains
0.0001 . . . . . . . . . . . . . . . . . . . . . . . . .square dekameters or ares
0.000395367 . . . . . . . . . . . . . . . . . . . . . . . . . . . . . . .square rods
0.01 . . . . . . . . . . . . . . . . . . . . . . . . . .square meters or centares
0.011960 . . . . . . . . . . . . . . . . . . . . . . . . . . . . . . . . square yards
0.0139498 . . . . . . . . . . . . . . . . . . . . . . . . . .square varas (Texas)
0.107639 . . . . . . . . . . . . . . . . . . . . . . . . . . . . . . . . . square feet
0.247104 . . . . . . . . . . . . . . . . . . . . . . . . . . . . . . . .square links
1 . . . . . . . . . . . . . . . . . . . . . . . . . . . . . . . . . . square decimeters
100 . . . . . . . . . . . . . . . . . . . . . . . . . . . . . . .square centimeters
15.5 . . . . . . . . . . . . . . . . . . . . . . . . . . . . . . . . . . .square inches
10,000 . . . . . . . . . . . . . . . . . . . . . . . . . . . . . . square millimeters
15,500,000 . . . . . . . . . . . . . . . . . . . . . . . . . . . . . . . . square mils
19,735,000 . . . . . . . . . . . . . . . . . . . . . . . . . . . . . . . circular mils
12,732 . . . . . . . . . . . . . . . . . . . . . . . . . . . . . circular millimeters

**SQUARE DEKAMETER: =**

0.00003831 . . . . . . . . . . . . . . . . . . . . . . . . . .square miles or sections
0.0001 . . . . . . . . . . . . . . . . . . . . . . . . . . . . . . . . square kilometers
0.00247104 . . . . . . . . . . . . . . . . . . . . . . . . . . . . . .square furlongs
0.0247104 . . . . . . . . . . . . . . . . . . . . . . . . . . . . . . . . . . . . .acres
0.01 . . . . . . . . . . . . . . . . . . . . . . . square hektometers or hectares
0.247104 . . . . . . . . . . . . . . . . . . . . . . . . . . . . . . . . .square chains
1 . . . . . . . . . . . . . . . . . . . . . . . . . . . . . . .square dekameters or ares
3.95367 . . . . . . . . . . . . . . . . . . . . . . . . . . . . . . . . . . . square rods
100 . . . . . . . . . . . . . . . . . . . . . . . . . . . . .square meters or centares
119.60 . . . . . . . . . . . . . . . . . . . . . . . . . . . . . . . . . . . square yards
139.498 . . . . . . . . . . . . . . . . . . . . . . . . . . . . .square varas (Texas)
1,076,39 . . . . . . . . . . . . . . . . . . . . . . . . . . . . . . . . . . . square feet
2,471.04 . . . . . . . . . . . . . . . . . . . . . . . . . . . . . . . . . .square links
10,000 . . . . . . . . . . . . . . . . . . . . . . . . . . . . . . . square decimeters
1,000,000 . . . . . . . . . . . . . . . . . . . . . . . . . . . . .square centimeters
155,000 . . . . . . . . . . . . . . . . . . . . . . . . . . . . . . . . . .square inches
100,000,000 . . . . . . . . . . . . . . . . . . . . . . . . . . . square millimeters
155,000,000,000 . . . . . . . . . . . . . . . . . . . . . . . . . . . . . square mils
197,350,000,000 . . . . . . . . . . . . . . . . . . . . . . . . . . . . . circular mils
127,320,000 . . . . . . . . . . . . . . . . . . . . . . . . . . . circular millimeters

## SQUARE FOOT: =

| | |
|---|---|
| 0.0000000355913 | square miles or sections |
| 0.0000000929034 | square kilometers |
| 0.00000229568 | square furlongs |
| 0.0000229568 | acres |
| 0.00000929034 | square hektometers or hectares |
| 0.000229568 | square chains |
| 0.000929034 | square dekameters or ares |
| 0.00367309 | square rods |
| 0.0929034 | square meters or centares |
| 0.111111 | square yards |
| 0.129598 | square varas (Texas) |
| 1 | square feet |
| 2.29568 | square links |
| 9.290341 | square decimeters |
| 929.0341 | square centimeters |
| 144 | square inches |
| 92,903.41152 | square millimeters |
| 144,000,000 | square mils |
| 183,346,560 | circular mils |
| 118,285 | circular millimeters |

## SQUARE FURLONG: =

| | |
|---|---|
| 0.015625 | square miles or sections |
| 0.040687 | square kilometers |
| 1 | square furlongs |
| 10 | acres |
| 4.04687 | square hektometers or hectares |
| 100 | square chains |
| 404.6873 | square dekameters or ares |
| 1,600 | square rods |
| 40,468.73 | square meters or centares |
| 48,400 | square yards |
| 56,453.069 | square varas (Texas) |
| 435,600 | square feet |
| 1,000,000 | square links |
| 4,046,873 | square decimeters |
| 404,687,300 | square centimeters |
| 62,726,500 | square inches |
| 40,468,730,000 | square millimeters |
| $627,265 \times 10^8$ | square mils |
| $79,865,038,655 \times 10^3$ | circular mils |
| 51,524,787,000 | circular millimeters |

SQUARE HEKTOMETER: =

| | |
|---|---|
| 0.003831 | square miles or sections |
| 0.01 | square kilometers |
| 0.247104 | square furlongs |
| 2.471044 | acres |
| 1 | square hektometers or hectares |
| 24.71044 | square chains |
| 100 | square dekameters or ares |
| 395.367 | square rods |
| 10,000 | square meters or centares |
| 11,959.8 | square yards |
| 13,949.8 | square varas (Texas) |
| 107,639 | square feet |
| 247,104 | square links |
| 1,000,000 | square decimeters |
| 100,000,000 | square centimeters |
| 15,500,000 | square inches |
| $1 \times 10^{10}$ | square millimeters |
| $155 \times 10^{11}$ | square mils |
| $19,735 \times 10^{8}$ | circular mils |
| $12,732 \times 10^{6}$ | circular millimeter |

SQUARE INCH: =

| | |
|---|---|
| 0.00000000024908 | square miles or sections |
| 0.000000000645163 | square kilometers |
| 0.0000000159422 | square furlongs |
| 0.000000159422 | acres |
| 0.0000000645163 | square hektometers or hectares |
| 0.00000159423 | square chains |
| 0.00000645163 | square dekameters or ares |
| 0.0000255076 | square rods |
| 0.000645163 | square meters or centares |
| 0.000771605 | square yards |
| 0.000899986 | square yards (Texas) |
| 0.00694444 | square feet |
| 0.0159423 | square links |
| 0.0645163 | square decimeters |
| 6.451626 | square centimeters |
| 1 | square inches |
| 645.16258 | square millimeters |
| 1,000,000 | square mils |
| 1,273,240 | circular mils |
| 821.423611 | circular millimeters |

## SQUARE KILOMETER: =

| | |
|---|---|
| 0.383101 | square miles or sections |
| 1 | square kilometers |
| 24.71044 | square furlongs |
| 247.1044 | acres |
| 100 | square hektometers or hectares |
| 2,471.044 | square chains |
| 10,000 | square dekameters or ares |
| 39,536.7 | square rods |
| 1,000,000 | square meters or centares |
| 1,195,980 | square yards |
| 1,394,980 | square varas (Texas) |
| 10,764,000 | square feet |
| 24,710,440 | square links |
| 100,000,000 | square decimeters |
| $1 \times 10^{10}$ | square centimeters |
| 1,550,000,000 | square inches |
| $1 \times 10^{12}$ | square millimeters |
| $155 \times 10^{13}$ | square mils |
| $19,735 \times 10^{10}$ | circular mils |
| $12,732 \times 10^{8}$ | circular millimeters |

## SQUARE LINK: =

| | |
|---|---|
| 0.000000015625 | square miles or sections |
| 0.0000000404687 | square kilometers |
| 0.000001 | square furlongs |
| 0.00001 | acre |
| 0.00000404687 | square hektometers or hectares |
| 0.0001 | square chains |
| 0.000404687 | square dekameters or ares |
| 0.0016 | square rods |
| 0.040469 | square meters or centares |
| 0.0484 | square yards |
| 0.0564531 | square varas (Texas) |
| 0.4356 | square feet |
| 1 | square links |
| 4.046873 | square decimeters |
| 404.6873 | square centimeters |
| 62.7265 | square inches |
| 40,468.73 | square millimeters |
| 62,726,500 | square mils |
| 79,863,380 | circular mils |
| 51,524.787 | circular millimeters |

## SQUARE METER: =

|  |  |
|---|---|
| 0.0000003831 | square miles or sections |
| 0.000001 | square kilometers |
| 0.0000247104 | square furlongs |
| 0.000247104 | acres |
| 0.0001 | square hektometers or hectares |
| 0.00247104 | square chains |
| 0.01 | square dekameters or ares |
| 0.0395367 | square rods |
| 1 | square meters or centares |
| 1.19598 | square yards |
| 1.39498 | square varas (Texas) |
| 10.7639 | square feet |
| 19.13580 | square spans |
| 24.71044 | square links |
| 96.8750 | square hands |
| 100 | square decimeters |
| 10,000 | square centimeters |
| 1,550 | square inches |
| 1,000,000 | square millimeters |
| 1,550,000,000 | square mils |
| 197,350,000 | circular mils |
| 1,273,200 | circular millimeters |

## SQUARE MIL: =

|  |  |
|---|---|
| 0.00000000000000024908 | square miles or sections |
| 0.00000000000000064516 | square kilometers |
| 0.0000000000000159422 | square furlongs |
| 0.000000000000159422 | acres |
| 0.0000000000000064516 | square hektometers or hectares |
| 0.00000000000159423 | square chains |
| 0.00000000000645163 | square dekameters or ares |
| 0.0000000000255076 | square rods |
| 0.000000000645163 | square meters or centares |
| 0.000000000771605 | square yards |
| 0.000000000899986 | square varas (Texas) |
| 0.00000000694444 | square feet |
| 0.0000000159423 | square links |
| 0.0000000645163 | square decimeters |
| 0.00000645163 | square centimeters |
| 0.000001 | square inches |
| 0.000645163 | square millimeters |
| 1 | square mils |
| 1.273224 | circular mils |
| 0.000821424 | circular millimeters |

## SQUARE MILE OR SECTION: =

| | |
|---|---|
| 1 | square miles or sections |
| 2.589998 | square kilometers |
| 64 | square furlongs |
| 640 | acres |
| 258.9998 | square hektometers or hectares |
| 6,400 | square chains |
| 25,899.98 | square dekameters or ares |
| 102,400 | square rods |
| 2,589,998 | square meters or centares |
| 3,097,600 | square yards |
| 3,612,995 | square varas (Texas) |
| 27,878,400 | square feet |
| 64,000,000 | square links |
| 25,899,980 | square decimeters |
| 258,999,800 | square centimeters |
| 4,014,496,900 | square inches |
| 2,589,998,000 | square millimeters |
| $40,144,969 \times 10^8$ | square mils |
| $51,114 \times 10^{10}$ | circular mils |
| $32,976 \times 10^8$ | circular millimeters |

## SQUARE MILLIMETER: =

| | |
|---|---|
| 0.0000000000003831 | square miles or sections |
| 0.000000000001 | square kilometers |
| 0.0000000000247104 | square furlongs |
| 0.000000000247104 | acres |
| 0.0000000001 | square hektometers or hectares |
| 0.0000000024104 | square chains |
| 0.00000001 | square dekameters or ares |
| 0.0000000395367 | square rods |
| 0.000001 | square meters or centares |
| 0.0000011960 | square yards |
| 0.00000139498 | square varas (Texas) |
| 0.0000107639 | square feet |
| 0.0000247104 | square links |
| 0.0001 | square decimeters |
| 0.01 | square centimeters |
| 0.00155 | square inches |
| 1 | square millimeters |
| 1,550 | square mils |
| 1,973.5 | circular mils |
| 1.2732 | circular millimeters |

SQUARE ROD: =

| | |
|---|---|
| 0.00000976563 | square miles or sections |
| 0.0000252929 | square kilometers |
| 0.000625 | square furlongs |
| 0.00625 | acres |
| 0.00252929 | square hektometers or hectares |
| 0.0625 | square chains |
| 0.252929 | square dekameters or ares |
| 1 | square rods |
| 25.2929 | square meters or centares |
| 30.25 | square yards |
| 35.283168 | square varas (Texas) |
| 272.25 | square feet |
| 625 | squrare links |
| 2,529.29 | square decimeters |
| 252,929 | square centimeters |
| 39,204 | square inches |
| 25,292,900 | square millimeters |
| 39,202,600,000 | square mils |
| 49,913,691,182 | circular mils |
| 32,202,992 | circular millimeters |

SQUARE VARA (TEXAS): =

| | |
|---|---|
| 0.000000274622 | square miles or sections |
| 0.000000716843 | square kilometers |
| 0.0000177135 | square furlongs |
| 0.000177135 | acres |
| 0.0000716843 | square hektometers or hectares |
| 0.00177135 | square chains |
| 0.00716843 | square dekameters or ares |
| 0.0283416 | square rods |
| 0.716843 | square meters or centares |
| 0.857332 | square yards |
| 1 | square varas (Texas) |
| 7.716 | square feet |
| 17.713467 | square links |
| 71.684263 | square decimeters |
| 7,168.426344 | square centimeters |
| 1,111.104 | square inches |
| 716,843 | square millimeters |
| 1,111,104,000 | square mils |
| 1,414,702,057 | circular mils |
| 912,687 | circular millimeters |

SQUARE YARD: =

| | |
|---|---|
| 0.000000322831 | square miles or sections |
| 0.000000836131 | square kilometers |
| 0.0000206612 | square furlongs |
| 0.000206612 | acres |
| 0.0000836131 | square hektometers or hectares |
| 0.00206612 | square chains |
| 0.00836131 | square dekameters or ares |
| 0.0330579 | square rods |
| 0.836131 | square meters or centares |
| 1 | square yards |
| 1.166382 | square varas (Texas) |
| 9 | square feet |
| 20.66112 | square links |
| 83.61306 | square decimeters |
| 8,361.306 | square centimeters |
| 1,296 | square inches |
| 836,131 | square millimeters |
| 1,296,000,000 | square mils |
| 1,650,119,040 | circular mils |
| 1,064,565 | circular millimeters |

TEMPERATURE, ABSOLUTE IN CENTIGRADE OR KELVIN: =

temperature in C° + 273.18°

TEMPERATURE, ABSOLUTE IN FAHRENHEIT OR RANKIN: =

temperature in F° + 459.69°

TEMPERATURE, DEGREES CENTIGRADE: =

5/9 (Temp. F° – 32°)
5/4 (Temp. Reaumur)

TEMPERATURE, DEGREES FAHRENHEIT: =

9/5 (Temp. C° + 32°)
9/4 (Temp. Reaumur + 32°)

**TEMPERATURE, DEGREES REAUMUR: =**

4/9 (Temp. F° – 32°)
4/5 (Temp. C°)

**Degree Centigrade: =**

.8 or 4/5 degree Reaumur
1.00 degrees absolute, Kelvin
1.8 or 9/5 degrees Fahrenheit

**Degree Fahrenheit: =**

0.44444 or 4/9 degree Reaumur
0.55556 or 5/9 degree Centigrade

**Degree Reaumur: =**

1.25 or 5/4 degrees Centigrade
2.25 or 9/4 degrees Fahrenheit

**TONS (LONG): =**

| | |
|---|---|
| 1 | ιons (long) |
| 1.0160470 | tons (metric) |
| 1.12 | tons (net) |
| 1,016.0470 | kilograms |
| 2,722.22 | pounds (Troy) |
| 2,240 | pounds (Avoir.) |
| 10,160.470 | hektograms |
| 101,605 | dekagrams |
| 32,667 | ounces (Troy) |
| 35,840 | ounces (Avoir.) |
| 1,016,047 | grams |
| 10,160,470 | decigrams |
| 101,604,700 | centigrams |
| 1,016,047,000 | milligrams |
| 15,680,000 | grains |
| 261,333 | drams (Troy) |
| 573,440 | drams (Avoir.) |
| 653,333 | pennyweights |
| 784,022 | scruples |
| 5,080,320 | carats (metric) |

TONS (LONG) (cont'd): =

| | |
|---|---|
| 6.19755 | barrels of water @ 60° F. |
| 7.33627 | barrels of oil @ 36° API |
| 28.607 | cubic feet |
| 260.02971 | gallons (U.S.—liquid) |

TONS (METRIC): =

| | |
|---|---|
| 0.984206 | tons (long) |
| 1 | tons (metric) |
| 1.10231 | tons (net) |
| 1,000 | kilograms |
| 2,679.23 | pounds (Troy) |
| 2,204.622341 | pounds (Avoir.) |
| 10,000 | hektograms |
| 100,000 | dekagrams |
| 32,150.76 | ounces (Troy) |
| 35,273.96 | ounces (Avoir.) |
| 1,000,000 | grams |
| 10,000,000 | decigrams |
| 100,000,000 | centigrams |
| 1,000,000,000 | milligrams |
| 15,432,365 | grains |
| 257,206 | drams (Troy) |
| 564,384 | drams (Avoir.) |
| 643,015 | pennyweights |
| 771,618 | scruples |
| 5,000,086 | carats (metric) |
| 6.297 | barrels of water @ 60° F. |
| 7.454 | barrels of oil @ 36° API |
| 29.0662 | cubic feet |
| 264.474 | gallons (U.S.—liquid) |

TONS (NET): =

| | |
|---|---|
| 0.892858 | tons (long) |
| 0.907185 | tons (metric) |
| 1 | tons (net) |
| 907.184872 | kilograms |
| 2,430.56 | pounds (Troy) |
| 2,000 | pounds (Avoir.) |
| 9,071.84872 | hektograms |
| 90,718.4872 | dekagrams |

TONS (NET) (cont'd): =

29,166.66 . . . . . . . . . . . . . . . . . . . . . . . . . . . . . . ounces (Troy)
32,000 . . . . . . . . . . . . . . . . . . . . . . . . . . . . . .ounces (Avoir.)
907,185 . . . . . . . . . . . . . . . . . . . . . . . . . . . . . . . . . grams
9,071,849 . . . . . . . . . . . . . . . . . . . . . . . . . . . . . . . decigrams
90,718,487 . . . . . . . . . . . . . . . . . . . . . . . . . . . . . centigrams
907,184,872 . . . . . . . . . . . . . . . . . . . . . . . . . . . .milligrams
14,000,000 . . . . . . . . . . . . . . . . . . . . . . . . . . . . . . . . grains
233,333 . . . . . . . . . . . . . . . . . . . . . . . . . . . . . . . drams (Troy)
512,000 . . . . . . . . . . . . . . . . . . . . . . . . . . . . . . drams (Avoir.)
583,333 . . . . . . . . . . . . . . . . . . . . . . . . . . . . . .pennyweights
700,020 . . . . . . . . . . . . . . . . . . . . . . . . . . . . . . . . . scruples
4,536,000 . . . . . . . . . . . . . . . . . . . . . . . . . . . . carats (metric)
5.71255 . . . . . . . . . . . . . . . . . . . . . . . . . . barrels of water @ 60° F.
6.76216 . . . . . . . . . . . . . . . . . . . . . . . . . . .barrels of oil @ 36° API
32.04 . . . . . . . . . . . . . . . . . . . . . . . . . . . . . . . . . . cubic feet
239.9271 . . . . . . . . . . . . . . . . . . . . . . . . . . gallons (U.S.—liquid)

TON OF REFRIGERATION: =

7,208,640,000 . . . . . . . . . . . . . . . . . . . . . . . . .foot poundals per day
300,360,000 . . . . . . . . . . . . . . . . . . . . . . . . . foot poundals per hour
5,006,000 . . . . . . . . . . . . . . . . . . . . . . . . . .foot poundals per minute
83,433.334 . . . . . . . . . . . . . . . . . . . . . . . . .foot poundals per second
224,056,200 . . . . . . . . . . . . . . . . . . . . . . . . . . .foot pounds per day
9,335,672 . . . . . . . . . . . . . . . . . . . . . . . . . . . . foot pounds per hour
15,594.53 . . . . . . . . . . . . . . . . . . . . . . . . . . .foot pounds per minute
2,593.242 . . . . . . . . . . . . . . . . . . . . . . . . . . . foot pounds per second
72,570.8 . . . . . . . . . . . . . . . . . . . . . . . . . . kilogram calories per day
3,023,784 . . . . . . . . . . . . . . . . . . . . . . . . . kilogram calories per hour
50.396 . . . . . . . . . . . . . . . . . . . . . . . . . . kilogram calories per minute
0.83994 . . . . . . . . . . . . . . . . . . . . . . . . . kilogram calories per second
2,559,836 . . . . . . . . . . . . . . . . . . . . . . . . . . . .ounce calories per day
107,060 . . . . . . . . . . . . . . . . . . . . . . . . . . . . . ounce calories per hour
1,777.664 . . . . . . . . . . . . . . . . . . . . . . . . . . .ounce calories per minute
29.62780 . . . . . . . . . . . . . . . . . . . . . . . . . . . .ounce calories per second
30,977,400 . . . . . . . . . . . . . . . . . . . . . . . . . . kilogram meters per day
1,290,722 . . . . . . . . . . . . . . . . . . . . . . . . . . kilogram meters per hour
21,512 . . . . . . . . . . . . . . . . . . . . . . . . . . kilogram meters per minute
358.534 . . . . . . . . . . . . . . . . . . . . . . . . . .kilogram meters per second
2,996,006 . . . . . . . . . . . . . . . . . . . . . . . . . . . liter-atmospheres per day
124,834 . . . . . . . . . . . . . . . . . . . . . . . . . . . liter-atmospheres per hour
2,080.554 . . . . . . . . . . . . . . . . . . . . . . . . .liter-atmospheres per minute
34.676 . . . . . . . . . . . . . . . . . . . . . . . . . . liter-atmospheres per second

TON OF REFRIGERATION (cont'd): =

| | |
|---|---|
| 1,058,693,800 | cubic foot atmospheres per day |
| 44,112,200 | cubic foot atmospheres per hour |
| 735,200 | cubic foot atmospheres per minute |
| 12,253.4 | cubic foot atmospheres per second |
| 4.655 | Cheval-Vapeur hours |
| 4.715 | horsepowers |
| 3.514 | kilowatts |
| 3,514 | watts |
| 303,609,600 | joules per day |
| 12,650,400 | joules per hour |
| 210,840 | joules per minute |
| 3,514 | joules per second |
| 19.7286 | pounds of carbon oxidized with 100% efficiency per day |
| 0.82202 | pounds of carbon oxidized with 100% efficiency per hour |
| 0.0137 | pounds of carbon oxidized with 100% efficiency per minute |
| 0.00022834 | pounds of carbon oxidized with 100% efficiency per second |
| 296.64 | pounds of water evaporated from and at 212° F. per day |
| 12.36 | pounds of water evaporated from and at 212° F. per hour |
| 0.206 | pounds of water evaporated from and at 212° F. per minute |
| 0.0034334 | pounds of water evaporated from and at 212° F. per second |
| 288,000 | BTU per day |
| 12,000 | BTU per hour |
| 200 | BTU per minute |
| 3.3334 | BTU per second |

TONS (NET) OF WATER PER DAY: =

| | |
|---|---|
| 0.0907226 | kiloliters per day |
| 0.00378023 | kiloliters per hour |
| 0.0000630030 | kiloliters per minute |
| 0.00000105004 | kiloliters per second |
| 0.0907226 | cubic meters per day |
| 0.00378023 | cubic meters per hour |
| 0.0000630030 | cubic meters per minute |
| 0.00000105004 | cubic meters per second |

## TON (NET) OF WATER PER DAY (cont'd): =

|  |  |
|---|---|
| 1.186579 | cubic yards per day |
| 0.0494404 | cubic yards per hour |
| 0.000824014 | cubic yards per minute |
| 0.0000137335 | cubic yards per second |
| 5.706411 | barrels per day |
| 0.237766 | barrels per hour |
| 0.00396285 | barrels per minute |
| 0.0000660467 | barrels per second |
| 0.907226 | hektoliters per day |
| 0.0378023 | hektoliters per hour |
| 0.00630030 | hektoliters per minute |
| 0.0000105004 | hektoliters per second |
| 32.0385859 | cubic feet per day |
| 1.334931 | cubic feet per hour |
| 0.0222490 | cubic feet per minute |
| 0.000370809 | cubic feet per second |
| 9.0722588 | dekaliters per day |
| 0.378023 | dekaliters per hour |
| 0.00630030 | dekaliters per minute |
| 0.000105004 | dekaliters per second |
| 239.664469 | gallons (U.S.) per day |
| 9.986099 | gallons (U.S.) per hour |
| 0.166433 | gallons (U.S.) per minute |
| 0.00277388 | gallons (U.S.) per second |
| 199.563810 | gallons (Imperial) per day |
| 8.315159 | gallons (Imperial) per hour |
| 0.138586 | gallons (Imperial) per minute |
| 0.00230977 | gallons (Imperial) per second |
| 90.722588 | liters per day |
| 3.780228 | liters per hour |
| 0.0630030 | liters per minute |
| 0.00105004 | liters per second |
| 90.722588 | cubic decimeters per day |
| 3.780228 | cubic decimeters per hour |
| 0.0630030 | cubic decimeters per minute |
| 0.00105004 | cubic decimeters per second |
| 958.657876 | quarts per day |
| 39.944877 | quarts per hour |
| 0.665740 | quarts per minute |
| 0.011095 | quarts per second |
| 1,917.315752 | pints per day |
| 79.888076 | pints per hour |
| 1.331470 | pints per minute |
| 0.0221913 | pints per second |

**TON (NET) OF WATER PER DAY (cont'd): =**

|  |  |
|---|---|
| 7,669.263008 | gills per day |
| 319.551826 | gills per hour |
| 5.325824 | gills per minute |
| 0.0887645 | gills per second |
| 907.225881 | deciliters per day |
| 37.802277 | deciliters per hour |
| 0.630030 | deciliters per minute |
| 0.0105004 | deciliters per second |
| 55,362.492339 | cubic inches per day |
| 2,306.770514 | cubic inches per hour |
| 38.446974 | cubic inches per minute |
| 0.640767 | cubic inches per second |
| 9,072.258810 | centiliters per day |
| 378.0227670 | centiliters per hour |
| 6.300300 | centiliters per minute |
| 0.105004 | centiliters per second |
| 90,722.588095 | milliliters per day |
| 3,780.227670 | milliliters per hour |
| 63.00299561 | milliliters per minute |
| 1.0500419 | milliliters per second |
| 90,722.588095 | cubic centimeters per day |
| 3,780.227670 | cubic centimeters per hour |
| 63.00299561 | cubic centimeters per minute |
| 1.0500419 | cubic centimeters per second |
| 90,722,588 | cubic millimeters per day |
| 3,780,227 | cubic millimeters per hour |
| 63,002.995611 | cubic millimeters per minute |
| 1,050.0419380 | cubic millimeters per second |
| 2,000 | pounds (Avoir.) per day |
| 83.333333 | pounds (Avoir.) per hour |
| 1.388889 | pounds (Avoir.) per minute |
| 0.0231481 | pounds (Avoir.) per second |
| 25.744757 | bushels (U.S.—dry) per day |
| 1.0726902 | bushels (U.S.—dry) per hour |
| 0.0178783 | bushels (U.S.—dry) per minute |
| 0.000297975 | bushels (U.S.—dry) per second |
| 24.958658 | bushels (Imperial) per day |
| 1.0399521 | bushels (Imperial) per hour |
| 0.0173323 | bushels (Imperial) per minute |
| 0.000288868 | bushels (Imperial) per second |

## TONS (NET) OF WATER PER HOUR: =

| | |
|---|---|
| 2.177342 | kiloliters per day |
| 0.0907226 | kiloliters per hour |
| 0.00151204 | kiloliters per minute |
| 0.0000252007 | kiloliters per second |
| 2.177342 | cubic meters per day |
| 0.0907226 | cubic meters per hour |
| 0.00151204 | cubic meters per minute |
| 0.0000252007 | cubic meters per second |
| 28.476932 | cubic yards per day |
| 1.186579 | cubic yards per hour |
| 0.0197763 | cubic yards per minute |
| 0.000329611 | cubic yards per second |
| 136.953864 | barrels per day |
| 5.706411 | barrels per hour |
| 0.0951061 | barrels per minute |
| 0.00158512 | barrels per second |
| 21.773421 | hektoliters per day |
| 0.907226 | hektoliters per hour |
| 0.0151204 | hektoliters per minute |
| 0.000252007 | hektoliters per second |
| 768.925102 | cubic feet per day |
| 32.0385859 | cubic feet per hour |
| 0.533972 | cubic feet per minute |
| 0.00889970 | cubic feet per second |
| 217.734211 | dekaliters per day |
| 9.0722588 | dekaliters per hour |
| 0.151204 | dekaliters per minute |
| 0.00252007 | dekaliters per second |
| 5,751.947256 | gallons (U.S.) per day |
| 239.664469 | gallons (U.S.) per hour |
| 3.994488 | gallons (U.S.) per minute |
| 0.0665740 | gallons (U.S.) per second |
| 4,789.531441 | gallons (Imperial) per day |
| 199.56381 | gallons (Imperial) per hour |
| 3.326064 | gallons (Imperial) per minute |
| 0.0554344 | gallons (Imperial) per second |
| 2,177.342114 | liters per day |
| 90.722588 | liters per hour |
| 1.512043 | liters per minute |
| 0.0252007 | liters per second |
| 2,177.342114 | cubic decimeters per day |
| 907.225881 | cubic decimeters per hour |
| 1.512043 | cubic decimeters per minute |
| 0.0252007 | cubic decimeters per second |

TONS (NET) OF WATER PER HOUR (cont'd): =

| | |
|---|---|
| 23,007.789024 | quarts (liquid) per day |
| 958.657876 | quarts (liquid) per hour |
| 15.977711 | quarts (liquid) per minute |
| 0.266291 | quarts (liquid) per second |
| 46,015.578048 | pints (liquid) per day |
| 1,917.315752 | pints (liquid) per hour |
| 31.954464 | pints (liquid) per minute |
| 0.532282 | pints (liquid) per second |
| 184,062 | gills (liquid) per day |
| 7669.263008 | gills (liquid) per hour |
| 127.820251 | gills (liquid) per minute |
| 2.130353 | gills (liquid) per second |
| 21,773.421143 | deciliters per day |
| 9,072.258810 | deciliters per hour |
| 15.120431 | deciliters per minute |
| 0.252007 | deciliters per second |
| 1,328,700 | cubic inches per day |
| 55,362.492339 | cubic inches per hour |
| 922.708206 | cubic inches per minute |
| 15.378550 | cubic inches per second |
| 217,734 | centiliters per day |
| 90,722.588095 | centiliters per hour |
| 151.204313 | centiliters per minute |
| 2.520072 | centiliters per second |
| 2,177,342 | milliliters per day |
| 907,226 | milliliters per hour |
| 1,512.0431349 | milliliters per minute |
| 25.200719 | milliliters per second |
| 2,177,342 | cubic centimeters per day |
| 907,226 | cubic centimeters per hour |
| 1,512.0431349 | cubic centimeters per minute |
| 25.200719 | cubic centimeters per second |
| 2,177,342,114 | cubic millimeters per day |
| 907,225,881 | cubic millimeters per hour |
| 15,120.431349 | cubic millimeters per minute |
| 25,200.718915 | cubic millimeters per second |
| 48,000 | pounds (Avoir.) per day |
| 2,000 | pounds (Avoir.) per hour |
| 33.333333 | pounds (Avoir.) per minute |
| 0.555556 | pounds (Avoir.) per second |
| 617.874174 | bushels (U.S.—dry) per day |
| 25.744757 | bushels (U.S.—dry) per hour |
| 0.429071 | bushels (U.S.—dry) per minute |
| 0.00715135 | bushels (U.S.—dry) per second |

TONS (NET) OF WATER PER HOUR (cont'd): =

| | |
|---|---|
| 599.00778724 | bushels (Imperial) per day |
| 24.958658 | bushels (Imperial) per hour |
| 0.415986 | bushels (Imperial) per minute |
| 0.00693301 | bushels (Imperial) per second |

TONS (NET) OF WATER PER MINUTE: =

| | |
|---|---|
| 1,306.415814 | kiloliters per day |
| 54.434991 | kiloliters per hour |
| 0.907226 | kiloliters per minute |
| 0.0151207 | kiloliters per second |
| 1,306.415814 | cubic meters per day |
| 54.434991 | cubic meters per hour |
| 0.907226 | cubic meters per minute |
| 0.0151207 | cubic meters per second |
| 1,708.673452 | cubic yards per day |
| 71.194727 | cubic yards per hour |
| 1.186579 | cubic yards per minute |
| 0.0197764 | cubic yards per second |
| 8,217.231850 | barrels per day |
| 342.384660 | barrels per hour |
| 57.0641101 | barrels per minute |
| 0.0951061 | barrels per second |
| 13,064.158138 | hektoliters per day |
| 544.349908 | hektoliters per hour |
| 9.0722588 | hektoliters per minute |
| 0.151207 | hektoliters per second |
| 46,135.563668 | cubic feet per day |
| 1,922.315153 | cubic feet per hour |
| 32.0385859 | cubic feet per minute |
| 0.533972 | cubic feet per second |
| 130,642 | dekaliters per day |
| 5,443.499084 | dekaliters per hour |
| 90.722588 | dekaliters per minute |
| 1.512067 | dekaliters per second |
| 345,117 | gallons (U.S.) per day |
| 14,379.86814 | gallons (U.S.) per hour |
| 239.664469 | gallons (U.S.) per minute |
| 3.994488 | gallons (U.S.) per second |
| 287,372 | gallons (Imperial) per day |
| 11,973.828603 | gallons (Imperial) per hour |
| 199.563810 | gallons (Imperial) per minute |
| 3.326064 | gallons (Imperial) per second |

## TONS (NET) OF WATER PER MINUTE (cont'd): =

| | |
|---|---|
| 1,306,416 | liters per day |
| 54,434.990844 | liters per hour |
| 907.225881 | liters per minute |
| 15.120671 | liters per second |
| 1,306,416 | cubic decimeters per day |
| 54,434.990844 | cubic decimeters per hour |
| 907.225881 | cubic decimeters per minute |
| 151.206710 | cubic decimeters per second |
| 1,380,467 | quarts per day |
| 57,519.472560 | quarts per hour |
| 958.657876 | quarts per minute |
| 15.977711 | quarts per second |
| 2,760,935 | pints per day |
| 115,039 | pints per hour |
| 1,917.315752 | pints per minute |
| 31.954464 | pints per second |
| 11,043,739 | gills per day |
| 460,156 | gills per hour |
| 7,669.263008 | gills per minute |
| 127.820970 | gills per second |
| 13,064,158 | deciliters per day |
| 544,350 | deciliters per hour |
| 9,072.258810 | deciliters per minute |
| 1,512.067104 | deciliters per second |
| 79,721,989 | cubic inches per day |
| 3,321,750 | cubic inches per hour |
| 55,362.492339 | cubic inches per minute |
| 922.708206 | cubic inches per second |
| 130,641,581 | centiliters per day |
| 5,443,499 | centiliters per hour |
| 90,722.588095 | centiliters per minute |
| 15,120.671014 | centiliters per second |
| 1,306,415,814 | milliliters per day |
| 54,434,991 | milliliters per hour |
| 907,226 | milliliters per minute |
| 151,207 | milliliters per second |
| 1,306,415,814 | cubic centimeters per day |
| 54,434,991 | cubic centimeters per hour |
| 907,226 | cubic centimeters per minute |
| 151,207 | cubic centimeters per second |
| $1,306,415,814 \times 10^3$ | cubic millimeters per day |
| 54,434,991,000 | cubic millimeters per hour |
| 907,226,000 | cubic millimeters per minute |
| 151,207,000 | cubic millimeters per second |

TON (NET) OF WATER PER MINUTE (cont'd): =

| | |
|---|---|
| 2,880,000 | pounds (Avoir.) per day |
| 120,000 | pounds (Avoir.) per hour |
| 2,000 | pounds (Avoir.) per minute |
| 33.333333 | pounds (Avoir.) per second |
| 37,072.450454 | bushels (U.S.—dry) per day |
| 1,544.685436 | bushels (U.S.—dry) per hour |
| 25.744757 | bushels (U.S.—dry) per minute |
| 0.429071 | bushels (U.S.—dry) per second |
| 35,940.467234 | bushels (Imperial) per day |
| 1,497.519468 | bushels (Imperial) per hour |
| 24.958658 | bushels (Imperial) per minute |
| 0.415986 | bushels (Imperial) per second |

TONS (NET) OF WATER PER SECOND: =

| | |
|---|---|
| 783,849 | kiloliters per day |
| 3,266.0395345 | kiloliters per hour |
| 54.434991 | kiloliters per minute |
| 0.907226 | kiloliters per second |
| 783,849 | cubic meters per day |
| 3,266.0395345 | cubic meters per hour |
| 54.434991 | cubic meters per minute |
| 0.907226 | cubic meters per second |
| 102,520 | cubic yards per day |
| 4,271.683630 | cubic yards per hour |
| 71.194727 | cubic yards per minute |
| 1.186579 | cubic yards per second |
| 493,034 | barrles per day |
| 20,543.0796248 | barrels per hour |
| 342.384660 | barrels per minute |
| 5.706411 | barrels per second |
| 783,849 | hektoliters per day |
| 32,660.390552 | hektoliters per hour |
| 544.340322 | hektoliters per minute |
| 9.0722588 | hektoliters per second |
| 2,768,134 | cubic feet per day |
| 115,339 | cubic feet per hour |
| 1,922.315153 | cubic feet per minute |
| 32.0385859 | cubic feet per second |
| 7,838,494 | dekaliters per day |
| 326,604 | dekaliters per hour |
| 5,443.398425 | dekaliters per minute |
| 90.722588 | dekaliters per second |

TONS (NET) OF WATER PER SECOND (cont'd): =

|  |  |
|---|---|
| 20,707,010 | gallons (U.S.) per day |
| 862,792 | gallons (U.S.) per hour |
| 14.379868 | gallons (U.S.) per minute |
| 239.664469 | gallons (U.S.) per second |
| 17,242,313 | gallons (Imperial) per day |
| 718,430 | gallons (Imperial) per hour |
| 11,973.828603 | gallons (Imperial) per minute |
| 199.563810 | gallons (Imperial) per second |
| 78,384,937 | liters per day |
| 3,266,039 | liters per hour |
| 54,433.984253 | liters per minute |
| 907.225881 | liters per second |
| 78,384,937 | cubic decimeters per day |
| 3,266,039 | cubic decimeters per hour |
| 54,433.984253 | cubic decimeters per minute |
| 907.225881 | cubic decimeters per second |
| 82,828,040 | quarts (liquid) per day |
| 3,451,168 | quarts (liquid) per hour |
| 57,519.47256 | quarts (liquid) per minute |
| 958.657876 | quarts (liquid) per second |
| 165,656,081 | pints (liquid) per day |
| 6,902,337 | pints (liquid) per hour |
| 115,039 | pints (liquid) per minute |
| 1,917.315752 | pints (liquid) per second |
| 662,624,324 | gills (liquid) per day |
| 27,609.347 | gills (liquid) per hour |
| 460,156 | gills (liquid) per minute |
| 7,669.263008 | gills (liquid) per second |
| 783,849,373 | deciliters per day |
| 32,660,391 | deciliters per hour |
| 544,340 | deciliters per minute |
| 9,072.258810 | deciliters per second |
| 4,783,319,338 | cubic inches per day |
| 199,304,972 | cubic inches per hour |
| 3,321,750 | cubic inches per minute |
| 55,362.492339 | cubic inches per second |
| 7,838,493,732 | centiliters per day |
| 326,603,906 | centiliters per hour |
| 5,443,398 | centiliters per minute |
| 90,722.588095 | centiliters per second |
| 78,384,937,325 | milliliters per day |
| 3,266,039,055 | milliliters per hour |
| 54,433,984 | milliliters per minute |
| 907,226 | milliliters per second |

TONS (NET) OF WATER PER SECOND (cont'd): =

| | |
|---|---|
| 78,384,937,325 | cubic centimeters per day |
| 3,266,039,055 | cubic centimeters per hour |
| 54,433,984 | cubic centimeters per minute |
| 907,226 | cubic centimeters per second |
| $78,384,937,325 \times 10^3$ | cubic millimeters per day |
| $3,266,039,055 \times 10^3$ | cubic millimeters per hour |
| 54,433,984,000 | cubic millimeters per minute |
| 907,226,000 | cubic millimeters per second |
| 172,800,000 | pounds (Avoir.) per day |
| 7,200,000 | pounds (Avoir.) per hour |
| 120,000 | pounds (Avoir.) per minute |
| 2,000 | pounds (Avoir.) per second |
| 2,224,347 | bushels (U.S.—dry) per day |
| 92,681.126136 | bushels (U.S.—dry) per hour |
| 1,544.685436 | bushels (U.S.) (dry) per minute |
| 25.744757 | bushels (U.S.—dry) per second |
| 2,156,428 | bushels (Imperial) per day |
| 89,851.168086 | bushels (Imperial) per hour |
| 150.605153 | bushels (Imperial) per minute |
| 24.958658 | bushels (Imperial) per second |
| 1,544.685436 | bushels (U.S.)dry) per minute |
| 25.744757 | bushels (U.S.—dry) per second |
| 2,156,428 | bushels (Imperial) per day |
| 89,851.168086 | bushels (Imperial) per hour |
| 150.605153 | bushels (Imperial) per minute |
| 24.958658 | bushels (Imperial) per second |

VARA (TEXAS): =

| | |
|---|---|
| 0.000456840 | miles (nautical) |
| 0.000526094 | miles (statute) |
| 0.000846670 | kilometers |
| 0.00420875 | furlongs |
| 0.00846670 | hektometers |
| 0.0420875 | chains |
| 0.0846670 | dekameters |
| 0.168350 | rods |
| 0.846670 | meters |
| 0.925926 | yards |
| 1 | varas (Texas) |
| 2.777778 | feet |
| 3.703704 | spans |
| 4.208754 | links |

VARA (TEXAS) (cont'd): =

| | |
|---|---|
| 8.333333 | hands |
| 8.466700 | decimeters |
| 84.667003 | centimeters |
| 33.333333 | inches |
| 846.670032 | millimeters |
| 33,333.333333 | mils |
| 846,670 | microns |
| 846,670,032 | milli-microns |
| 846,670,032 | micro-millimeters |
| 1,315,011 | wave lengths of red line of cadmium |
| 8,466,700,319 | Angstrom units |

WATT: =

| | |
|---|---|
| 63,725.184 | foot pounds per day |
| 2,655.22 | foot pounds per hour |
| 44.2536 | foot pounds per minute |
| 0.73756 | foot pounds per second |
| 764.702 | inch pounds per day |
| 31,862.5920 | inch pounds per hour |
| 531.04320 | inch pounds per minute |
| 8.85072 | inch pounds per second |
| 20.640096 | kilogram calories (mean) per day |
| 0.860004 | kilogram calories (mean) per hour |
| 0.0143334 | kilogram calories (mean) per minute |
| 0.00023889 | kilogram calories (mean) per second |
| 45.503616 | pound calories (mean) per day |
| 1.895984 | pounds calories (mean) per hour |
| 0.0315997 | pound calories (mean) per minute |
| 0.000526662 | pound calories (mean) per second |
| 728.0578547 | ounce calories (mean) per day |
| 30.335744 | ounce calories (mean) per hour |
| 0.505596 | ounce calories (mean) per minute |
| 0.00842659 | ounce calories (mean) per second |
| 20,640.096 | gram calories (mean) per day |
| 860.004 | gram calories (mean) per hour |
| 14.3334 | gram calories (mean) per minute |
| 0.23889 | gram calories (mean) per second |
| 81.930528 | BTU (mean) per day |
| 3.413772 | BTU (mean) per hour |
| 0.056896 | BTU (mean) per minute |
| 0.00094827 | BTU (mean) per second |
| 0.001 | kilowatts |

WATT (cont'd): =

1 . . . . . . . . . . . . . . . . . . . . . . . . . . . . . . . . . . . . . watts
3,600 . . . . . . . . . . . . . . . . . . . . . . . . . . . . . . . . . . joules
0.001341 . . . . . . . . . . . . . . . . . . . . . . . . . . . horsepowers
0.0013597 . . . . . . . . . . . . . . . . . . . . . horsepowers (metric)
0.0013597 . . . . . . . . . . . . . . . . . . . . . Cheval-Vapeur hours
0.000234 . . . . . . . . . . . . . pounds carbon oxidized with 100% efficiency
0.00352 . . . . . . . . . . . . . . . . . . . . . . . . pounds water evaporated from and at 212° F.
0.852647 . . . . . . . . . . . . . . . . . . . . . kiloliter-atmospheres per day
0.0355270 . . . . . . . . . . . . . . . . . . . kiloliter-atmospheres per hour
0.000592116 . . . . . . . . . . . . . . . . . kiloliter-atmospheres per minute
0.0000098686 . . . . . . . . . . . . . . . . kiloliter-atmospheres per second
852.647 . . . . . . . . . . . . . . . . . . . . . . . . liter-atmospheres per day
35.52695 . . . . . . . . . . . . . . . . . . . . . . . liter-atmospheres per hour
0.592116 . . . . . . . . . . . . . . . . . . . . . . liter-atmospheres per minute
0.0098686 . . . . . . . . . . . . . . . . . . . . . liter-atmospheres per second
8,808 . . . . . . . . . . . . . . . . . . . . . . . . . . kilogram meters per day
367.1 . . . . . . . . . . . . . . . . . . . . . . . . . . kilogram meters per hour
6.116667 . . . . . . . . . . . . . . . . . . . . . kilogram meters per minute
0.101944 . . . . . . . . . . . . . . . . . . . . . . kilogram meters per second

WATT HOUR: =

2,655.22 . . . . . . . . . . . . . . . . . . . . . . . . . . . . . . . . foot pounds
31,982.64 . . . . . . . . . . . . . . . . . . . . . . . . . . . . . . . inch pounds
0.860004 . . . . . . . . . . . . . . . . . . . . . . kilogram calories (mean)
1.895984 . . . . . . . . . . . . . . . . . . . . . . . . pound calories (mean)
30.385744 . . . . . . . . . . . . . . . . . . . . . . . ounce calories (mean)
860.004 . . . . . . . . . . . . . . . . . . . . . . . . . gram calories (mean)
3.413772 . . . . . . . . . . . . . . . . . . . . . . . . . . . . . BTU (mean)
0.001 . . . . . . . . . . . . . . . . . . . . . . . . . . . . . kilowatt hours
3,600 . . . . . . . . . . . . . . . . . . . . . . . . . . . . . . . . . . joules
0.001341 . . . . . . . . . . . . . . . . . . . . . . . . . horsepower hours
0.0013597 . . . . . . . . . . . . . . . . . . . . horsepower hours (metric)
0.0355270 . . . . . . . . . . . . . . . . . . . . . . . . kiloliter-atmospheres
35.53695 . . . . . . . . . . . . . . . . . . . . . . . . . . . liter-atmospheres
367.1 . . . . . . . . . . . . . . . . . . . . . . . . . . . . . . kilogram meters

WATT PER SQUARE INCH: =

8.1913 . . . . . . . . . . . . . . . . . . . . BTU per square foot per minute
6,372.6 . . . . . . . . . . . . . . . . foot pounds per square foot per minute
0.19310 . . . . . . . . . . . . . . . . . . . . . . . horsepowers per square foot

YARD: =

| | |
|---|---|
| 0.000493387 | miles (nautical) |
| 0.000568182 | miles (statute) |
| 0.000914404 | kilometers |
| 0.00454545 | furlongs |
| 0.00914404 | hektometers |
| 0.0454545 | chains |
| 0.0914404 | dekameters |
| 0.181818 | rods |
| 0.914404 | meters |
| 1 | yards |
| 1.08 | varas (Texas) |
| 3 | feet |
| 4 | spans |
| 4.54545 | links |
| 9 | hands |
| 9.144036 | decimeters |
| 91.440360 | centimeters |
| 36 | inches |
| 914.40360 | millimeters |
| 36,000 | mils |
| 914,404 | microns |
| 914,403,600 | milli-microns |
| 914,403,600 | micro-millimeters |
| 1,420,212 | wave lengths of red line of cadmium |
| 9,144.036,345 | Angstrom units |

# PERSONAL CONVERSION FACTORS